Lecture Notes in Earth Sciences 57

Editors:
S. Bhattacharji, Brooklyn
G. M. Friedman, Brooklyn and Troy
H. J. Neugebauer, Bonn
A. Seilacher, Tuebingen and Yale

Springer
Berlin
Heidelberg
New York
Barcelona
Budapest
Hong Kong
London
Milan
Paris
Tokyo

Elisabeth Lallier-Vergès
Nicolas-Pierre Tribovillard
Philippe Bertrand

Organic Matter Accumulation

The Organic Cyclicities of the Kimmeridge Clay Formation (Yorkshire, GB) and the Recent Maar Sediments (Lac du Bouchet, France)

Springer

Authors

Dr. Elisabeth Lallier-Vergès
U.R.P.O., Laboratoire de Géologie de la Matière Organique
Université d'Orléans
BP 6759, F-45067 Orléans Cedex 2, France

Dr. Nicolas-Pierre Tribovillard
Université Paris Sud, Sciences de la Terre
Bâtiment 504, F-91405 Orsay Cedex, France

Dr. Philippe Bertrand
Université Bordeaux I, Géologie et Océanologie
Avenue des Facultés, F-33405 Talence Cedex, France

"For all Lecture Notes in Earth Sciences published till now please see final pages of the book"

ISBN 3-540-59170-2 Springer-Verlag Berlin Heidelberg New York

Library of Congress Cataloging-in-Publication Data.

Organic matter accumulation: the organic cyclities of the Kimmeridge Clay Formation (Yorkshire, GB) and the recent maar sediments (Lac du Bouchet, France) / [edited by] Elisabeth Lallier-Vergès, Nicolas-Pierre Tribovillard, Philippe Bertrand.
p.cm. – Lecture notes in earth sciences: 57)
ISBN 0-387-59170-2. – ISBN 3-540-59170-2
1. Sedimentation and deposition – England – Yorkshire. 2. Organic geochemistry – England – Yorkshire. 3. Sedimentation and deposition – France – Bouchet Lake. 5. Kimmeridge Clay (England and Scotland) I. Lallier-Vergès, Elisabeth. II. Tribovillard, Nicolas-Pierre, 1962- III. Bertrand, Philippe, 1954- . IV. Series
QE571.0736 1995 552'.5'094281–dc20 95-12984

Printed in Germany

Typesetting: Camera ready by author
SPIN: 10492885 32/3142-543210 - Printed on acid-free paper

Preface

The four-year period of activity of the *Groupement de Recherche 942* (GDR) of the *Centre National de la Recherche Scientifique* (CNRS) came to an end in December 1993. This GDR was a scientific association grouping research teams from the academic sphere — i.e. the *Unités de Recherches Associées* 723 & 724 of the CNRS as well as the Universities of Orléans and Paris-Sud — and from the industrial world: Elf-Aquitaine Production, TOTAL and the Institut Français du Pétrole (IFP).

The aim of the GDR was to understand the processes and the causes of organic carbon fossilization in sediments, especially when they can be modified by environmental conditions such as climate, eustatism, productivity etc., factors which can also interact. This goal implies the simultaneous study of ancient geological formations (hydrocarbon source rocks from the famous Kimmeridge Clay Formation) and recent Quaternary sediments (the Lac du Bouchet or lake Bouchet maar, Massif Central, France). In the latter case, we benefit from a fine-scale stratigraphical framework as well as a reliable reconstruction of the local and regional environment.

This volume is a collection of papers representing oral presentations given on December 7, 1993, at the Société Géologique de France in Paris, during the final meeting of the GDR. These articles thus report the latest developments of the studies carried out under the GDR. However, this is not the first publication of our results, which can be found in the papers referred to in each article.

The Kimmeridge Clay Formation was previously studied in 1987, by the Yorkim Group from IFP, Elf-Aquitaine and the British Geological Survey, on the basis of a series of wells drilled across the Cleveland Basin of Yorkshire. In each well, the distribution with depth of the total organic content is cyclic. We have compared some of the organic cycles from two wells (Marton and Ebberston) based on mineralogy, organic and inorganic geochemistry and petrography, at a high resolution scale (centimetric). The main conclusion of this work is that the driving force for organic matter accumulation in the studied cycles was organic phytoplankton productivity. Oxygenation conditions seem to have played a secondary role as a positive feedback action enhancing organic matter storage.

Lac du Bouchet is located on the Devès volcanic plateau, 15 km SW of Le Puy en Velay, at an altitude of 1205 m. The depth of the water column is 28 m. The lake has a subcircular shape (1 km in diameter) and a very restricted watershed. This site is exceptionally suitable for research on climate variations and palaeomagnetic field modifications (Euromaars EC Program). The GDR focused on sedimentary organic matter and its relationship to inorganic phases. An important result is that organic matter appears to be a good indicator of palaeoenvironmental reconstructions for over 350 000 years. In addition, the study of early diagenetic reactions in surficial sediments (porewater and solid phase) allows the specification of the processes of organic matter degradation and storage in such an oligothrophic lake.

Acknowledgements Elf-Aquitaine Production, TOTAL and the Institut Français du Pétrole (IFP), as well as the Centre National de la Recherche Scientifique and the Universities of Orléans and Paris Sud, are thanked for their scientific contributions and financial support. Eugène Bonifay and Nicolas Thouveny from the *Laboratoire de Géologie du Quaternaire* (CNRS, Marseille, France) and the European Program EUROMAARS are greatly thanked for their help during fieldwork, sample processing and scientific discussions. We are also indebted to Elizabeth Jolivet, Andrew J. Patience and Simo Boussafir (University of Orléans) for improving the final manuscript.

Table of contents-

The Organic Cyclicities of the Kimmeridge Clay Formation (Yorkshire, UK)

E. Lallier-Vergès, P. Bertrand, N.-P. Tribovillard and A. Desprairies
Short-Term Organic Cyclicities from the Kimmeridge Clay Formation of Yorkshire (UK): Combined Accumulation and Degradation of Organic Carbon under the Control of Primary Production Variations 3

M. Boussafir, E. Lallier-Vergès, P. Bertrand and D. Badaut-Trauth
SEM and TEM Studies on Isolated Organic Matter and Rock Microfacies from a Short-Term Organic Cycle of the Kimmeridge Clay Formation (Yorkshire, UK) 15

F. Gelin, M. Boussafir, S. Derenne, C. Largeau and P. Bertrand
Study of Qualitative and Quantitative Variations in Kerogen Chemical Structure Along a Microcycle: Correlation with Ultrastructural Features 31

J.-R. Disnar and L. Ramanampisoa
Palaeoproduction Control on Anoxia and Organic Matter Preservation and Accumulation in the Kimmeridge Clay Formation of Yorkshire (UK): Molecular Assessment 49

A. Desprairies, M. Bachaoui, A. Ramdani and N.-P. Tribovillard
Clay Diagenesis in Organic-Rich Cycles from the Kimmeridge Clay Formation of Yorkshire (UK): Implication for Palaeoclimate interpretations 63

The Recent Maar Sediments (Lac du Bouchet, France)

E. Viollier, P. Albéric, M. Evrard, D. Jézéquel, D. Lavergne, G. Michard, M. Pèpe, G. Sarazin and P. Zuddas
Geochemical Study of the Lac du Bouchet, Haute-Loire, France. Part I: Water Balance and Biogeochemical Implications 95

D. Jézéquel, P. Albéric, A. Desprairies, M. Evrard, D. Lavergne, G. Michard, A.J. Patience, M. Pepe, G. Sarazin, N.-P. Tribovillard and E. Viollier
Geochemical Study of the Lac du Bouchet, Haute-Loire, France. Part II: Water-Sediment-Organic Matter Interactions during the Last 2 500 Years 119

A. J. Patience, E. Lallier-Vergès, A. Sifeddine, P. Albéric and B. Guillet
Organic Fluxes and Early Diagenesis in the Lacustrine Environment: the Superficial Sediments of thge Lac du Bouchet (Haute-Loire, France) 145

A. Sifeddine, P. Bertrand, E. Lallier-Vergès and A.J. Patience
Organic Sedimentation and its Relationship with Palaeoenvironmental Changes over the last 30 000 Years (Lac du Bouchet, Haute Loire, France). Comparison with Other Palaeoclimatic Lacustrine Examples........... 157

B. Guillet and Ousmane Maman
Sulphur Speciation in the Late Glacial and Holocene Sediments of the Lac du Bouchet (Haute Loire, France). 169

List of Contributors 183

Subject Index 185

THE ORGANIC CYCLICITIES OF THE KIMMERIDGE CLAY FORMATION (YORKSHIRE, GB)

Short-term organic cyclicities from the Kimmeridge Clay Formation of Yorkshire (G.B.): combined accumulation and degradation of organic carbon under the control of primary production variations

Elisabeth Lallier-Vergès[1] Philippe Bertrand,[1] Nicolas Tribovillard[2] and Alain Desprairies[2].

1) Université d'Orléans, URA 724 du CNRS, Dépt. des Sciences de la Terre, F-45067 Orléans cedex
2) Université Paris Sud, URA 723 du CNRS, bâtiment 504, F-91405 Orsay cedex

Key words- *Kimmeridge Clay Formation, organic carbon cyclicity, primary production, microbial degradation, modelisation.*

Abstract- This paper summarises different results acquired on the short-term organic cycles (about 30 000 years for one metre thick) from the Kimmeridge Clay Formation (Yorkshire, G.B.). The results, as a whole, support the fundamental interpretation that considers primary production as the main factor influencing organic cyclicity. Other developments specify that in addition to the high organic primary production, the consecutive microbial sulphate reduction, the intensity of which strictly depends on the organic-walled phytoplanktonic production, emphasises the accumulation of the produced hydrocarbon-rich organic matter. This specific accumulation is possible through the selective preservation of bio-resistant macro-molecules already present in living organisms and/or through early vulcanisation of lipidic molecules (Boussafir *et al.*; Gelin *et al.*, this volume). The latter is favoured by the low availibility of reduced iron species compared to the massively produced HS^- and consequently by the incorporation of HS^- in excess within the organic matter. Here, this interpretation is presented as a mathematical model whose results are compared to measured data (for microcycles having organic carbon contents ranging from about 2 to 9 %) and which account for the cyclic variation in both quantity and geochemical quality of organic matter.

Introduction-

The Kimmeridge Clay Formation is a marine deposit composed of alternating organic-rich shales and marls that is considered as the lateral equivalent of the main source-rocks of the North Sea. These immature formations outcrop on the south coast of England (i.e. Dorset) whereas others were drilled in the Cleveland basin (Yorkshire) by the YORKIM Group (Herbin and Geyssant, 1993). These deposits present cyclicities of several orders (Herbin et al., 1993; Desprairies *et al.*; Disnar and Ramanampisoa, this volume) which concern their organic content in terms of amount (total organic carbon content expressed as %TOC) and HC-potential (hydrogen index expressed as HI in mg HC/g C org)

The main objective of the GdR 942 research program was to answer the question: what is the nature and the origin of this organic carbon cyclicity? For this, short-term organic cycles representing about 30 000 years for one meter in thickness, have been sampled. One (CYCLE 1) is located at the base of a second-term organic cycle in the *Eudoxus* Zone

(Hole Marton 87) and records variations in TOC from about 2 to 9 %, the second (CYCLE 2) is situated at the peak of the same second-term organic cycle and exhibits TOC variations from about 4 to 30%. A high resolution study of these organic-rich rocks has been completed on both the mineral and the organic fractions of the rocks. The organic matter has been investigated by both petrographical and geochemical methods.

The aim of this paper is to propose a general interpretation based on some of the results of our research group and also from the litterature. Most of these studies (including their respective analytical techniques) are either presented in this volume or already published. They concern the characterisation of the mineral content of the organic cycles, the identification of the organic content and of its preservation state. Here, we specifically develop the importance of the microbial sulphate reduction in the close variation of both the quantity and the chemical quality of organic matter and we propose a mathematical model which accounts, at best, for the organic carbon cyclicity. The main idea is to propose a new highlight concerning the processes of accumulation of HC-enriched organic shales and to infer it to the source-rock deposition.

Result summary-

Composition of the sediment and nature of the organic matter-

The mineralogical and geochemical study of the bulk samples first indicates that the organic cyclicity is really due to the variation of organic matter content. No dilution due to the mineral phases occurs and no variation in the mineral composition has been found throughout the cycles (Tribovillard *et al.*, 1992). A study of the trace element behaviour shows that environmental depositional conditions were always anoxic, even if some differences in the O_2 depletion intensity may exist between the two cycles studied (Tribovillard *et al.*, 1994).

Petrographical studies have been completed on the organic matter isolated from the mineral matrix by acidic treatments and on the organic matter analysed *in situ* in the microtexture of the rock. These studies were performed by the means of optical studies (Ramanampisoa *et al.*, 1992; Pradier and Bertrand, 1992) and electron microscopy studies (cf. Boussafir *et al.* 1994; this volume). The sediment is always laminated and no bioturbation has been found, whatever the TOC, attesting the anoxic depositional conditions.

The composition of the organic matter varies with the TOC content. This variation mainly concerns the "orange amorphous organic matter" proportion (as described in palynofacies preparation by Ramanampisoa *et al.*, 1992) corresponding to the bituminite maceral observed on polished sections (Boussafir *et al*,. this volume). The amount of the "orange amorphous organic matter" increases when TOC increases, whereas the land-derived organic debris and the "brown amorphous organic matter" are mainly representative of the low TOC samples.

TEM investigations (cf. Lallier-Vergès *et al.*, 1993a; Boussafir *et al.* 1994, this volume) and pyrolitic analyses (cf. Gelin *et al.*, 1994, this volume) performed on the different types of organic matter has given the following results. The orange amorphous organic matter, nanoscopically amorphous when observed by transmission electron microscope (TEM) and always associated to pyrite framboids or crystals, is chemically composed of organo-sulphur compounds, whose the early formation may derive from the vulcanisation of phytoplankton-derived flocs. The brown amorphous organic matter, composed of ultralaminae when observed by TEM and without any associated sulphides, is chemically composed by lipidic macromolecules, thought to be derived from the selective preservation of the phytoplanktonic cell-walls.

The refractory or resistant character of the preserved organic matter is thus either, inherited (ultralaminae, land-derived debris); or acquired (amorphous organic matter) as reported by Lallier-Vergès *et al.* (1993a) and Boussafir *et al.* (1994). In both cases (excepted for lignaceous debris), the acquisition of this resistant feature is accompanied by an enrichment in hydrocarbon molecules. In low-TOC samples, when the lignaceous debris (very poor HC-content) proportion is high, the average HC-content of the bulk organic matter (i.e. HI) is inferior to that one of high-TOC samples, when the organic matter is mainly formed by hydrocarbon molecules.

Sulphate reduction process-

Whilst the sedimentary sulphide formation is related to the microbial reduction of the porewater sulphate, the sedimentary sulphur content is assumed to represent the metabolisable part of organic matter that has been microbially degraded in the anoxic domain. The study of the carbon and sulphur content evolution throughout the cycles accounts for the evolution of sulphate reduction intensity (i.e. the possible variations of degradation processes).

Figure 1 exhibits the positive relationship between sulphur content and TOC, as first defined by Berner and Raiswell (1983). Indeed, the higher the delivered amount of organic matter is, the higher the amount of metabolisable organic matter arriving at the water-sediment interface and the more intense the sulphate reduction process will be. As a consequence, the sedimentary sulphide content is positively correlated to the non-metabolised (resistant) organic matter content (TOC).

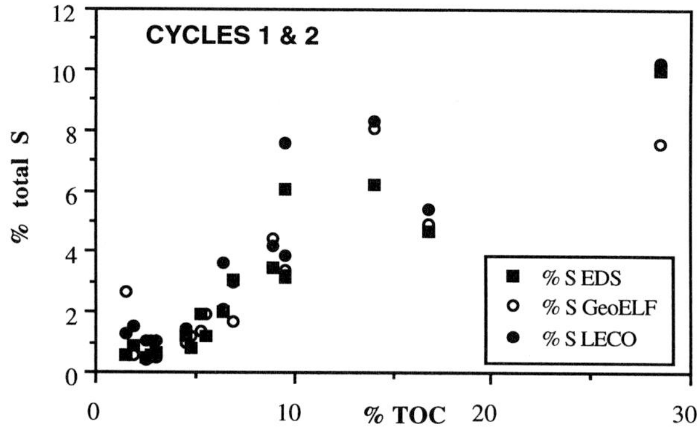

Figure 1 - Total sulphur content (dry weight %) analysed by Enery Dispersive X-ray Spectrometry on sample powders versus Total Organic Carbon content (dry weight %) obtained by Rock Eval Pyrolysis.

We determined the sulphate reduction index (SRI) such as already defined by Lallier-Vergès *et al.* 1993b and 1993c).

SRI = % initial organic carbon / % preserved organic carbon

Initial organic carbon is calculated as the sum of the preserved organic carbon (TOC) and the metabolised organic carbon (sulphur contents corrected from the stoichiometric sulphate reduction equation). This index is a minimum value considering the arbitrary assumption that no escape of HS^- occurs. On going studies based on the isotopical composition of sulphur taking into account the retention of HS^- confirm the reality of SRI evolution. The latter represents the variations of delivered organic matter in terms of metabolisable/resistant organic matter ratio. The figure 2 attests that this ratio changed throughout both the cycles.

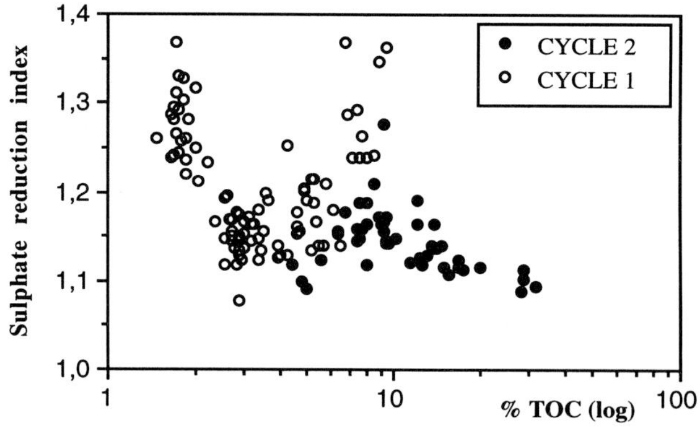

Figure 2 - The Sulphate Reduction Index as defined above *versus* the Total Organic Carbon content in dry weight % (log scale).

For low TOC values (< 3%), the SRI values increase when TOC values decrease. This indicates that nearly the whole available metabolisable organic matter has been microbially degraded. Petrographical studies (Boussafir *et al.*, this volume) show that the samples with the highest SRI have the greatest microfossile content (Foraminifers, ammonites, bivalves...), some of these skelettons being replaced by pyrite. These samples also contain pellets composed of coccolith tests. The occurrence of such grazer remains indicate a low primary productivity. Moreover, these species are known to be composed of mainly easily metabolisable organic matter (proteins, carbohydrates...) and a mineral skeleton. The degradation of this metabolisable organic matter in the anoxic domain will result in only very few sulphides which have no HC-enriched resistant organic matter associated. This may explain the high SRI values despite the low TOC and HI values encountered.

For TOC values ranging from 3 to 6%, the SRI values are almost constant showing that the metabolisable part associated with the resistant organic matter delivered from the photic zone was constant. This is represented in the S *versus* TOC diagram by a virtually linear relationship (fig. 1). That suggests an increasing productivity without any change in nature in the delivered organic matter.

Above a TOC value equal to 6%, the SRI increases. This trend shows a drastic change in the nature of the delivered organic matter. The amount of metabolisable organic matter associated with the resistant organic matter increased markedly with the rate of delivery of organic matter. Consequently, these samples which present a very high organic carbon accumulation, also record the greatest degradation of organic carbon. The threshold value (about 6% TOC) is also found in the study of the molecular evolution of bitumen throughout the cycles (Disnar and Ramanampisoa, this volume).

For TOC values greater than 10% (CYCLE 2), a decrease of the SRI is observed and interpreted as a progressive limitation of the sulphate reduction process, due to the progressive limitation of sulphate or other metabolites for the sulphate reducers (Calvert and Pedersen, 1992; Lallier-Vergès *et al.*, 1993b). This trend is also slightly visible on the S *versus* TOC diagram (fig. 1).

The speciation of sulphur (pyritic and organic sulphur) was performed by analysing the S, C and Fe contents of the residues obtained after HCl, and HF treatments of the bulk rock samples. When considering both the cycles studied, organic sulphur contents are directly proportional to the preserved organic carbon content (TOC), whereas those of pyrite are stabilised for the highest values of organic carbon (fig.3a and 3b). This trend has already been shown in other Kimmeridgian organic-rich rocks from Dorset outcrops (Lallier-Vergès *et al.*, 1993c).

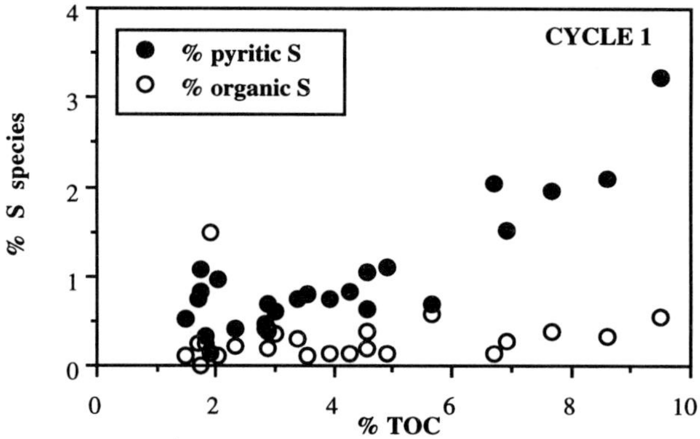

Figure 3a - % organic sulphur and % pyritic sulphur expressed versus TOC (dry weight %) for CYCLE 1 samples.

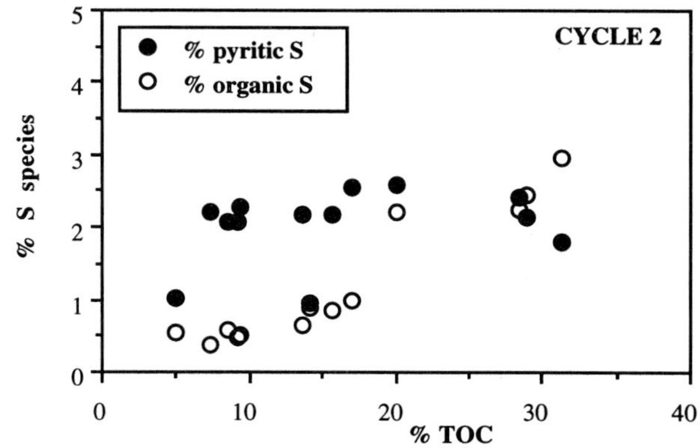

Figure 3b - % organic sulphur and % pyritic sulphur expressed versus TOC (dry weight %) for CYCLE 2 samples.

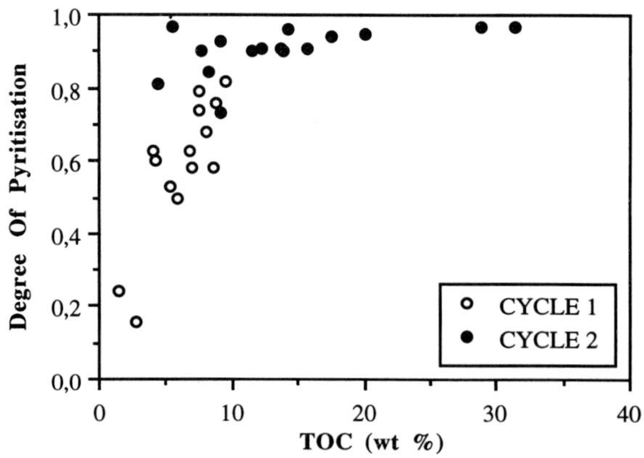

Figure 4: Degree of pyritisation (DOP) versus Total Organic Carbon (dry weight %) for the two cycles studied (after Tribovillard *et al.* 1994). The DOP consists of the ratio of reduced iron content /total reactive iron content as defined by Raiswell *et al.* (1988)

In the second cycle (fig.4), the available iron content is a limiting factor in pyrite formation, as is demonstrated by the study of the pyritisation degree (Tribovillard *et al.*, 1994) and HS^- in excess may be fixed in the organic matter (Lallier-Vergès *et al.*, 1994). Indeed, Boussafir *et al.*, (this volume) have shown that samples having the highest TOC and HI values are characterised by a nanoscopically amorphous organic matter which is notably enriched in sulphur.

Interpretative model and conclusion-

This model considers that the cyclicity is primarily controlled by production variations of mineral-free phytoplankton, whereas the benthic environment is always anoxic. Such variations would induce modifications in the rate of recycled organic carbon in the photic zone, so that the delivery rate of metabolisable carbon towards the bottom would be modified. Because the redox conditions were not strongly modified, this change may be attributed to a best yield of delivery out of the photic zone and thus to a least recycling of organic carbon in the photic zone. Biological processes, such as the formation of aggregates during algal blooms (Jackson, 1990), as well as the differential growing of phytoplanktonic and zooplanktonic populations, may explain the modifications in the yield of delivery (Wefer, 1989). The biological functioning of the photic zone could more precisely explain the variations of the sulphate reduction intensity compared to the organic fluxes recorded in the sediment. In particular, the marked increased degradation, corresponding the high organic contents, can be explained by a rough supply of mucilage aggregates during productive periods (Alldredge and Gotschalk, 1989), which would deliver a massive content of metabolisable organic matter (enriched in polysaccharides), favouring the sulphate reduction process.

The interpretative model is based on these conceptual hypotheses, which take into account the different results presented above and the recent advancements in the knowledge of the present biological functioning of the photic zone. The mathematical development of the model has already been presented (Bertrand and Lallier-Vergès, 1993; Bertrand *et al.*, 1994) and is not detailed here.

The figure 5 shows the good fit between so-calculated and measured values, for the relationship between the amount (TOC) and the HC-content (HI) of the preserved organic matter.

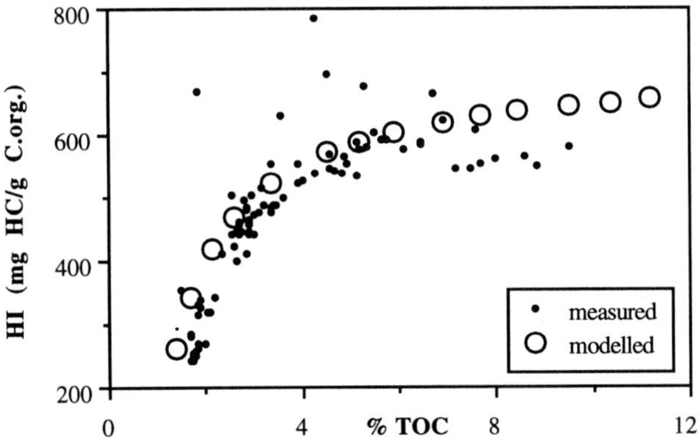

Figure 5: Comparison of calculated and measured values in the Hydrogen Index versus Total Organic Carbon content diagram for CYCLE 1 samples (after Bertrand and Lallier-Vergès, 1993).

The model also involves that the palaeofluxes of organic carbon, which have been mineralised through anaerobic diagenesis, are partially recorded by the sedimentary sulphide content. Figure 6 underlines that this model also accounts for the relationship between the sulphate reduction process and the organic matter preservation.

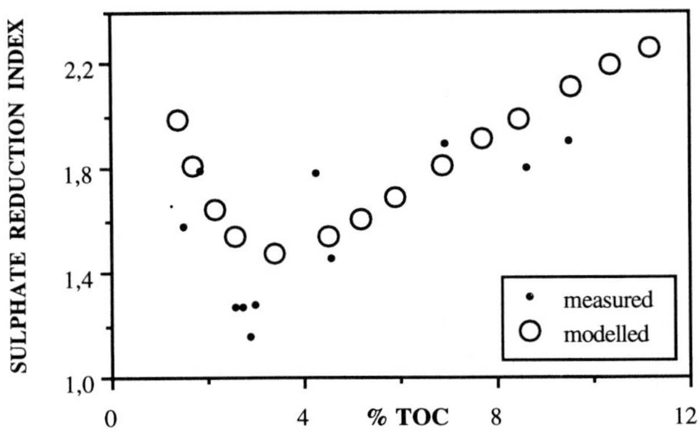

Figure 6: Comparison of calculated and measured values in the Sulphate Reduction Index (SRI) versus Total Organic Carbon (dry weight %) diagram for CYCLE 1 samples (after Bertrand and Lallier-Vergès, 1993).

We propose that high rates of HS^- production related to high rates of delivered metabolisable organic matter from the photic zone (due to phytoplankton-derived flocs as shown by Boussafir *et al.*, this volume) may favour the early formation of organo-sulphur compounds as already described by Sinninghe Damsté *et al.*, (1989). The additional molecular data are in agreement with this interpretation (Gelin *et al.*, 1994; this volume). The early vulcanisation of HC-enriched phytoplankton-derived organic matter would allow its preservation from any further microbial degradation.

The relationship between the quantity and the hydrocarbon content of the marine sedimentary organic matter was already presented for Kimmeridge rocks from Dorset by Huc *et al.*, (1992) and described as a dilution curve between low-HI land-derived organic matter and high-HI phytoplankton-derived organic matter. However, the authors thought that although this dilution factor is related to production variations, some degradation processes must be responsible for the drastic change in HI values for TOC contents < 10%. The present results suggest that such a trend may be explained by a sharp increase in the sulphate reduction intensity (due to a large delivery of metabolisable organic matter) which may dramatically increase the "vulcanisation process" and thus the hydrocarbon preservation. Biological processes, such as the development of planktonic populations, also present drastic trends, as is the case for the blooms.

It is therefore quite realistic to assume that a high accumulation of organic carbon may correspond, at the same time, to a lower preservation of organic carbon in term of fluxes and to a better chemical preservation of the residual organic matter, through both the processes of selective preservation and vulcanisation. The early vulcanisation of the phytoplankton-derived flocs would allow the long-term fossilisation of the aliphatic-enriched organic matter. Here, this phenomenon would be enhanced both by the occurrence of mineral-free phytoplankton and the limitation of available iron.

As proposed in Desprairies *et al.* (this volume), mineralogical indices show that palaeoclimatic changes have occurred throughout the micro-cycles, which implies mineral flux variations. Moreover, the results indicate that the highest mineral fluxes occurred at the same period as the organic fluxes, which is in agreement with the fact that the mineral input releases available nutrients for phytoplanktonic production. As a matter of fact, the limitation of available iron compared to HS^- formation, would be solely to the variation of the organic flux / mineral flux ratio.

Acknowledgements- We are indebted to our colleagues of the GdR 942 research group for their scientific contribution to this work.

References-

Alldredge A.L. and Gotschalk C.C. (1989) Direct observations of the mass flocculation of diatom blooms: characteristics, settling velocities and formation of diatom aggregates. *Deep-Sea Research*, **36**(2), 159-171.

Berner R. and Raiswell R. (1983) Burial of organic carbon and pyrite sulphur in sediments over Phanerozoic time: a new theory. *Geochim. Cosmochim. Acta,* **48**, 605-615.

Bertrand P. and Lallier-Vergès E. (1993). Past sedimentary organic matter accumulation and degradation controlled by productivity. *Nature*, **364**, 786-788.

Bertrand P., Lallier-Vergès E. and Boussafir M (1994). Enhancement of both accumulation and anoxic degradation of organic carbon controlled by cyclic productivity: a model. *Organic Geochemistry*, in press.

Boussafir M., Lallier-Vergès E., Bertrand P. and Badaut-Trauth D. (1994) Structure ultrafine de la matière organique des roches mères du kimméridgien du Yorkshire (UK). *Bull. Soc. Géol. Fr.*, **165**(4), 355-363.

Calvert S.E. and Pedersen T.F. (1992) Organic carbon preservation and accumulation in marine sediments: How important is the anoxia? *In*: J.Whelan and J. Farrington (Editors), Productivity, Accumulation, and Preservation of Organic Matter in Recent and Ancient Sediments. Columbia Univ. Press, New York, pp. 241-263.

Herbin J.P. and Geyssant J.R. (1993). "Ceintures organiques" au Kimméridgien/Tithonien en Angleterre (Yorkshire, Dorset) et en France (Boulonnais). *C.R. Acad. Sci. Paris*, **317**, 1309-1316.

Herbin J.P., Müller C., Geyssant J.R., Mélières F. and Penn I.E. (1993) Variation of the Distribution of organic matter within a transgressive system tract: Kimmeridge Clay (Jurassic), England. In: A.A.P.G.1993 "Petroleum source rocks in a sequence stratigraphic framework". B. KATZ and L. PRATT (eds). pp. 67-100.

Huc A.Y., Lallier-Vergès E., Bertrand P., Carpentier B. and Hollander D. J. (1992) Organic matter response to change of depositional environment in Kimmeridgian shales, DORSET, U.K. *In*: "Organic matter: productivity, accumulation and preservation of organic matter in recent and ancient sediments". J. WHELAN and J. FARRINGTON (eds). Columbia Univrsity Press. New York. pp. 469-486.

Jackson G.A. (1990) A model of the formation of marine algal flocs by physical coagulation processes. *Deep-Sea Research*, **37**(8), 1197-1211.

Lallier-Vergès E., Boussafir M., Bertrand P. and Badaut-Trauth D. (1993a) Selective preservation of various organic matter types as assessed by STEM studies on a cyclic productivity-controlled sedimentary series (Kimmeridge Clay Formation). *In* "Organic Geochemistry: Poster sessions from the 16th International Meeting on Organic Geochemistry, Stavanger, 1993". KJELL ØYGARD (ed.).pp. 384-386.

Lallier-Vergès E., Bertrand P. and Desprairies A. (1993b) Organic matter composition and sulfate reduction intensity in Oman Margin sediments. *Mar. Geol.*, **112**, 57-69.

Lallier-Vergès E., Bertrand P., Huc A.Y., Bückel D. and Tremblay P. (1993c) Control of the preservation of organic matter by productivity and sulphate reduction in Kimmeridgian shales from Dorset (UK). *Mar. and Petrol. Geol.*, **10**, 600-605.

Lallier-Vergès E., Bertrand P., Tribovillard N.P., Hayes J., Boussafir M., Zaback D.A. and Connan J. (1994) Productivity-induced sulfur enrichment of organic-rich sediments. 207th American chemical Society 1994, National Meeting - March 13-18, San Diego.

Gelin F., Derenne S., Largeau C. and Bertrand P. (1994) Study of Kimmeridge Clay Kerogen *via* combined petrographic and chemical methods. 207th American chemical Society 1994, National Meeting - March 13-18, San Diego.

Pradier B. and Bertrand P. (1992) Etude à haute résolution d'un cycle de carbone organique des argiles du Kimméridgien du Yorkshire (GB): relation entre composition pétrographique du contenu organique observé in-situ, teneur en carbone organique et qualité pétroligène. *C.R. Acad. Sc. Paris*, **315**(2), 187-192.

Raiswell R., Buckley F., Berner R.A. and Anderson T.F. (1988) Degree of pyritization as a palaeoenvironmental indicator of bottom water oxygenation. *J. Sediment. Petr.*, **58**, 812-819.

Ramanampisoa L., Bertrand P., Disnar J.R., Lallier-Vergès E., Pradier B. and Tribovillard N.P. (1992) Etude à haute résolution d'un cycle de carbone organique des argiles du Kimméridgien du Yorkshire (GB): résultats préliminaires de géochimie et de pétrographie organique. *C.R. Acad. Sc. Paris*, **314**(2), 1493-1498.

Sinninghe Damsté J.S., Rijpstra W.I., Kock-Van Dalen A.C., De Leeuw J.W. and Schenk P.A. (1989). Quenching of labile functionalised lipids by inorganic sulphur species: Evidence for the formation of sedimentary organic sulphur compounds at the early stages of diagenesis. *Geoch. et Cosmoch. Acta*, **55**, 1343-1355.

Tribovillard N.P., Desprairies A., Bertrand P., Lallier-Vergès E., Disnar J.R. and Pradier B. (1992) Etude à haute résolution d'un cycle du carbone organique de roches kimméridgiennes du Yorkshire (GB): minéralogie et géochimie (résultats préliminaires). *C. R. Acad. Sc.* Paris, **314**(2), 923-930.

Tribovillard N.P., Desprairies A., Lallier-Vergès E., Bertrand P., Moureau N., Ramdani A. and Ramanampisoa L. (1994) Geochemical study of organic matter rich cycles from the Kimmeridge Clay Formation of Yorkshire (UK): productivity versus anoxia. *Palaeogeogr., Palaeoclim., Palaeoecol.*, **108**, 165-181.

Wefer G. (1989) Particles flux in the Oceans: effects of episodic production. *In* Report of the Dahlem Workshop on Productivity of the Oceans: Present and Past. BERGER W.H., SMETACEK V.S. and WEFER G. (eds). Wiley, Chichester, 1989, pp. 139-154.

SEM and TEM studies on isolated organic matter and rock microfacies from a short-term organic cycle of the Kimmeridge Clay Formation (Yorkshire, G.B.)

Mohammed Boussafir[1], *Elisabeth Lallier-Vergès*[1], *Philippe Bertrand*[1] *and Denise Badaut-Trauth*[2]

1) Université d'Orléans, URA 724 du CNRS, Dépt. des Sciences de la Terre, F-45067 Orléans cedex
2) Museum d'Hist. Naturelle de Paris, URA 723 du CNRS, 53 rue Buffon, F-75005 Paris

Key-words- *Organic carbon cycle, ultrafine structure, amorphous organic matter, macerals, sulphate reduction, primary production.*

Abstract- Organic petrographical studies have been completed on short-term organic cyclicities of the Kimmeridge Clay Formation to assess the nature and the state of preservation of the organic content throughout the cycles. Optical investigations, performed on both the isolated organic matter and the organic material observed *in situ,* reveal a resistant organic matter mainly composed of optically amorphous material and land-derived debris. This amorphous organic matter is composed of two main types of ultrafine structures when observed by transmission electron microscopy (TEM). Each ultrafine structure corresponds to a specific degradation process. The "ultralaminae" derive from phytoplankton through a strong degradation before deposition and a selective preservation of the most resistant part of the plankton. The "nanoscopically amorphous organic matter" results from the delivery of phytoplankton-derived flocs, which reach the anoxic domain with a high content of metabolisable organic matter. The consecutive enhanced sulphate reduction liberates high contents of HS^- a part of which is incorporated in the organic matter. This process (vulcanisation) results in an amorphous organic material always associated with sulfides (mineral and organic). Its close association with clays and its planar distribution in the microtexture of the rock suggest a mode of deposition such as algal mats. A TEM study of *in situ* macerals allows to assess that the bituminite maceral corresponds to this material.

Introduction-

As a general rule, the sediments the richest in organic matter are derived from an environment in which a considerable biological primary productivity is coupled with conditions favourable for preservation. What are these conditions favourable for better preservation of the organic matter and, moreover, what are the phenomena which intervene between phytoplanktonic production and the final fossilisation of part of the original organic matter?

Using modern analytical techniques, one can describe the total organic matter, measure the organic carbon content (TOC) and the oil potential (hydrogen index: HI) of the rock, characterise and quantify the nature of the molecular composition and also the physico-chemical characteristics of the organic fraction.

To date, no detailed study of the individual organic constituents has been completed in order to understand their exact nature, origin and mode of fossilisation. This is mainly due to the fact that the normal techniques for isolating these organic components (from the rock), often result in fragments which are too small (average 30 μm) for reliable analysis.

This work forms part of a larger scientific framework (this volume) concerning the pre- and syn-diagenetic history of organic matter in short-term sedimentary cyclicity of the Kimmeridgian Clays of Yorkshire (Desprairies *et al.*, this volume). It is essentially based on the electron microscopy study of both organic matter isolated from the mineral matrix by acidic attacks and organic matter studied *in situ* in the microtexture of the rock. The scale of investigation ranges from the millimetre (optical microscopy) to the nanometre (transmission electron microscopy). For this, an organic carbon cycle (approximately 1.20 m) was sampled from Marton 87 cores and a multidisciplinary study was completed on these samples.

This cyclicity is marked by the variation of the quantity of organic matter (TOC) which varies from 1.8 to 9.5 %, associated with the geochemical quality (HI) which varies from 200 to 800 mg HC/g C-org. (Ramanampisoa *et al.*, 1992; Herbin *et al.*, 1993). The petrographical composition (nature and amount of the constituents) also varies along the cycle (Pradier and Bertrand, 1992; Ramanampisoa *et al.*, 1992).

The organic matter isolated from the mineral matrix-

Textural and microtextural features-

Optical analysis of the isolated organic matter (palynofacies) allowed identification of the organic components and evaluation of their proportion. The organic matter is mainly composed of three main types of amorphous organic matter (AOM) with constituents derived from the terrestial biomass (Pl. 1a) and zooplankton in small quantities.

The AOM was classed depending on its colour and texture. The most abundant AOM is represented by orange homogeneous flakes with sharp edges (orange AOM, Pl. 1b). AOM also occurs as brown heterogeneous flocs (brown AOM, Pl. 1c) and opaque aggregates (black AOM, Pl. 1d). Quantification of the palynofacies throughout the cycle illustrates the existence of a strong correspondence between the variations in TOC and HI and those of the orange AOM proportion (Ramanampisoa *et al.*, 1992).

Ultrastructural features-

Each AOM type was separated from the palynological preparation and concentrated using a micromanipulator coupled to a binocular. The obtained samples were fixed in OsO_4, embedded in resin and then prepared for observation under transmission electron microscope (TEM), according to the method described in Boussafir *et al.* (1994a). The results are the following.

The orange AOM has a nanoscopically amorphous (perfectly homogeneous) texture without any apparent structure even at high magnifications (Pl. 2a, 2b). The brown AOM is mainly composed of laminar structures (ultralaminae), more or less elongated and always with a regular thickness (50 to 400 nm), (Pl. 2c, 2d) which is generally greater than a cellular membrane (7.5 to 12 nm). These ultralaminar structures have already been observed by Raynaud *et al.* (1988), Largeau *et al.* (1990), Derenne *et al.* (1991; 1993), Lugardon *et al.* (1991) who consistently quote the thickness of these ultralaminations as being smaller than 60 nm. The black AOM contains several types of organic ultrafine structures always dominated by lignaceous debris (Pl. 2e). Associated with these terrestrial inclusions, one finds some granular organic matter (Pl. 2h), some small particles of nanoscopically amorphous organic matter (Pl. 2e) and some ultralaminar structures.

In order to assess the distribution throughout the carbon microcycle of the different organic ultrafine structures, eight representative samples of the microcycle were chosen to study ultra-thin sections of the bulk organic matter by TEM (Boussafir *et al.*, 1994b). Observation of the bulk organic matter allowed, in addition to the previously described ultrafine structures corresponding to the different AOM, the identification of other organic ultrafine stuctures, occurring in minor quantities and the observation of associated sulphides.

Finally, two categories of nanoscopically amorphous organic matter were distinguishable.

- The first one is composed of large massive homogeneous areas with a strong electron contrast. These zones, when compacted, form a homogeneous gel corresponding to the orange AOM. Within this nanoscopically amorphous organic matter, one can distinguish some anastomosed zones with discontinuous linear structures, that we named diffuse laminations (Pl. 2g).
- The second one is composed of less homogeneous regions of AOM which present a weaker electron contrast than the previous category and a speckled aspect (more or less granular) without any particular form (Pl. 2h).

Iron sulphides were found in diverse assemblages, always associated with the nanoscopically amorphous organic matter and the diffuse organic matter. Several forms have been distinguished and analysed by Energy Dispersive X-ray Spectrometry:

- rare isolated automorphous pyrite ranging from 0.5 to 1.5 µm (Pl. 3a)
- assemblage of automorphous pyrite microcrystals (<100 nm) (Pl. 3b).
- pyrite framboids associated with the nanoscopically AOM (Pl. 3c).
- and rare non-crystallised FeS spherules (Pl. 3d).

The observed ultrastructures have been grouped into four large groups and their semi-quantification shows that the samples at the peak of the cycle (highest TOC values), are also those which have the greatest proportion of pyrite and nanoscopically amorphous organic matter. However, the samples from the beginning and end of the cycle present low proportions of pyrite and AOM, and thus more abundant proportions of ultralaminations.

Chemical characterisation of some ultrastructural types of AOM-

Pin-point analyses were performed using energy-loss spectrometry (EELS) on a scanning transmission electron microscope (STEM) VG HB 501 coupled with a GATAN 666 spectrometer (Disko, 1992). EELS is preferable to energy dispersive X-ray spectrometry (EDXS) for analysing the light elements, such as: C, O, N which have a low fluorescence yield. Moreover, EELS technique offers more spatially localised analyses. This type of analysis was completed.on ultra-thin sections of organic matter, set on holed carbon-film grids. The results are presented under the form of atomic ratios (Table. 1). The nanoscopically amorphous organic matter *(nAOM)* from several levels throughout the cycle has been analysed as shown on the micrograph Pl. 3e.

Atomic Ratio	*homogeneous nAOM*	*homogeneous & diffuse nAOM*	*diffuse AOM*
N(S) / N(C)	0.11	0.09	0.05
N(N) / N(C)	0.03	0.04	0.02
N(O) / N(C)	0.06	0.02	0.04

Table 1: EELS analyses in nanoscopically amorphous organic matter.

The atomic ratios of sulphur over carbon "N(S) / N(C)" show that the more homogeneous and massive the organic matter, the greater the number of sulphur atoms relative to carbon atoms. Indeed, the N(S) / N(C) ratio is about 0.05 in the diffuse organic matter, about 0.09 in an intermediate organic matter and reaches about 0.11 in the homogeneous amorphous organic matter. This homogeneous organic matter also presents the highest O/C ratio (0.06).

The organic matter in the microtexture of the rock-

Microfacies-

Observation of the microfacies was performed by optical microscopy (reflected light and UV excitation on polished sections) and by scanning electron microscopy (SEM) with back-scattered electron imagery. As previously carried out on other samples from the KCF (Bertrand *et al.*, 1990; Belin and Brosse, 1992), this work was completed in order to illustrate the distribution of both the organic matter and the associated sulphides throughout the carbon cycle and in addition the evolution of the rock microtexture.

The microfacies are always laminated throughout the cycle, whatever the TOC value. No bioturbation features have been observed conversely to the microfacies samples studied in the Kimmeridge Clay from Dorset, South England (Bertrand *et al.*, 1990; Huc *et al.*, 1992).

The mineral fraction is largely dominated by clays. Silica is present as small smooth detrital grains (Pl. 4a, 4b) and diagenetic crystallisations including pyrite (Pl. 4b). Calcium carbonate is present in the form of planktonic tests. The coccoliths occur as fecal pellets (Pl. 4h) or dispersed in the organo-mineral groundmass (Pl. 4c). The foraminifers are often filled with pyrite crystals (Pl. 4d) and sometimes totally replaced by pyrite. Carbonates are also present in the form of automorphous calcitic and dolomitic diagenetic crystals (Pl. 4a, 4d).

The distribution of the organic matter was shown to be relatively heterogeneous. In samples with TOC values <4%, the organic matter is essentially made up of small irregular or angular particles, from 5 to 20 μm in size (Pl. 4a and 4b), defined as inertinite under reflected light. Some small and thin elongated organic particles are also present (Pl. 4g). For samples with TOC values >4%, microfacies are characterised by the increase in size and abundance of the elongated organic particles, more or less continuous (Pl. 4e). The higher the TOC value, the higher the frequency of these particles. This frequency reaches its maximum at the peak of the cycle. In these samples, the organic matter is present in two morphologically different types. One consists of fine elements roughly 30 μm in thickness which may be defined as alginite under reflected light (yellow-coloured under UV excitation, Pl. 5c). The other is represented by elongated particles thicker than the previous type (about 500 μm, Pl. 4e). When studied by reflected light, they appear as bituminite and present a brown colour under UV excitation (Pl. 5a). These particles have already been described and named "brown algae" by Pradier and Bertrand (1992). Moreover, the optical study allows the identification of *Tasmanacae algae* in small quantities all along the cycle (Pl. 5e). When the TOC values are high, the organic matter tends to form thick laminations (Pl. 4e) but also occurs within the mineral matrix in greater amounts (Pl. 4c). When studied on sections parallel to the stratification (Pl. 4f, 4h), the organic laminations do not show any particular form but are often elongated on surfaces of several hundreds of μm^2 in area. Pyrite framboids are always associated to these organic thick laminations (Pl. 4f).

Ultrafine structure of macerals-

TEM studies were completed on small fragments of the rocks, in order to study the ultrastructure of specific organic components and their relationship with the mineral groundmass. The petrographical components (macerals, organo-mineral matrix...) were selected under the microscope. The fragments obtained were prepared, with a procedure similar to that used for isolated organic matter, and then ultra-thin sections were performed and observed by TEM.

The internal structure of the bituminite particles appears as a nanoscopically amorphous and homogeneous organic matter (Pl. 5b), including pyrite framboids but also clay particles. This close association between clays and bituminite has already been described in other source-rock samples (Bishop *et al.*, 1994). As the same ultrastructure was observed in the concentrated orange AOM of the palynofacies preparation, it is therefore now possible to definitely correlate the bituminite liberated from mineral particles by acidic attack, with the orange amorphous organic matter, observed in palynofacies preparation.

The alginite particles have the same nanoscopically amorphous ultrastructure as bituminite, and are recognisable only by their filamentous form and small size (Pl. 5d). Nevertheless, these bodies keep their individual chemical characteristics, i.e. strong fluorescence under UV excitation (Pl. 5c).

TEM studies of *Tasmanacae* show that they consist of thick walls with a weaker contrast than the homogeneous AOM. The walls do not show any cellular or membranar organisation and have a gelified internal structure (Pl. 5f).

Thus, from the *in situ* study of organic matter, the optically identified macerals essentially consist of a nanoscopically amorphous ultrafine structure. Surprisingly, one notes the absence of any organised structure and, in particular, the ultralaminations found in the isolated organic matter. This is probably due to the fact that this material does not occur as a maceral and is difficult to select under the optical microscope.

Discussion-

Electron microscope studies of selected macerals and AOM have shown that bituminite as seen in polished section corresponds to the orange amorphous organic matter as seen in palynofacies preparation, and presents a nanoscopically amorphous ultrafine structure. This material is dominant in the levels richest in organic carbon. The ultrastructural studies of the total organic matter have underlined the role that sulphate reduction, which is more intense in the middle of the cycle than at the beginning or the end, has played in the generation of this nanoscopically amorphous organic matter.

It is known that sulphate reduction is one of the major degradation phenomena operating during early diagenesis of the organic matter in the anaerobic marine environment (Jørgensen, 1978, Henrichs and Reeburgh, 1987). One part of the H_2S, produced by the sulphate reducers, is fixed in the sediments in the form of ferrous sulfides when easily reducible iron is available (Tribovillard *et al.,* 1994).

Parallel studies have taken into account the sulphate reduction index of sediments (Lallier-Vergès *et al.*, 1992, 1993; this volume; Bertrand and Lallier-Vergès, 1993; Bertrand *et al.*, 1994) and shown that degradation by sulphate reducers was more intense in the levels richest in organic carbon (in terms of carbon balance). Moreover, it appears that when the iron becomes limiting, the sulphur is incorporated into the organic matter (Lallier-Vergès *et al.*, 1994).

It is shown here by EELS analyses that the sulphur in excess is preferentially fixed in the nanoscopically amorphous organic matter. The leading molecular studies of kerogen (Gelin *et al.*, this volume) indicate that the chemical nature of the sulphur fixation in the hydrogenated molecules could be the vulcanisation process.

From these results, we propose that this organic matter originates from the "gelification" of the remains of phytoplanktonic organic matter through microbial degradation. Alternatively, its planar texture could have been inherited from the mode of sediment deposition, and then emphasised by compaction. Moreover, it seems to act as a cement for the mineral particles.

Conversely, in the zones poor in organic matter, the clays cement the diffuse organic particles, which are mainly composed of bio-resistant material. There is therefore a sedimentological tie between the organic and mineral textures. The proposed hypothesis on the basis of these studies is that the bituminite precursor has been formed at the sediment-water interface and has included during its gelification the syn-sedimentary clay particles. This suggests a mode of deposition similar to that of algal mats. The zones enriched in bituminite would correspond to the periods of dominantly organic productivity as opposed to clay sedimentation.

Conclusion-

The detailed petrographical·study of a cyclic organic-rich deposit has thus added to the knowledge of the fossilisation process of organic carbon in the following way.

The accumulation of the bio-resistant organic matter (ultralaminations, lignaceous debris) in sediments having a low organic carbon content, responds to a selective preservation process of bio-macromolecules already existing in the living organisms. The parallel increase in the TOC and HI of an amorphous organic matter rich in pyrite and organic sulphur was in response to a selective preservation of the hydrogenated resistant macromolecules due to the sulphur incorporation. The succession of these processes and the evolution of sulphate reduction intensity throughout the cycle show that various amounts of metabolisable organic matter have been delivered to the water-sediment interface (Lallier-Vergès *et al.*, this volume). This confirms the hypothesis that the cyclicity of the organic carbon content in the Kimmeridge Clay of Yorkshire is mainly due to variations in biological productivity.

While productivity is in a steady state (i.e when the "grazers" recycle a large part of the primary productivity), the phytoplanktonic matter exported from the photic zone loses, during transfer in the water column, its hydrosoluble parts and/or easily degradable molecules, either due to the reaction with the oxygenated sea water, or from enzymatic degradation in the food chain. Only the resistant parts (bio-macromolecules) may reach the sea bed together with the lignaceous debris and the zooplankton remains (foraminifers...).

The bio-resistant organic matter would correspond to a weaker phytoplanktonic exportation from the photic zone (beginning and end of the cycle). Conversely, the inherited organic matter would correspond to a greater exportation from the photic zone linked to a strong phytoplanktonic production (peak of the cycle). Indeed, biological models show that when productivity is high, the grazers can no longer more recycle all the phytoplanktonic organic matter produced, and at that time, the phytoplankton may die in large quantities and form large flocs of phytodetritus (Jackson, 1990), enriched in polysaccharides. This flocculation process increases both their settling velocity and their metabolisable organic matter content (Wefer, 1989).

This ultrastructural study of HC-enriched organic matter from the Kimmeridge Clay Formation has therefore assessed both the nature of the organic matter present in these cyclicities and the fundamental role of sulphate reduction in the control of the quality (HC-content) of the preserved organic matter. The intensity of the sulphate reduction is, of course, primarily regulated by the production of organic phytoplankton, i.e. without mineral tests.

Acknowledgements- This work was funded by a study allowance from the "Conseil de la Région Centre" (France). We would like to thank Dominique Jalabert of the "Service Commun de Microscopie Electronique" (University of Orléans) and Annick Genty of the "Ecole Supérieure d'Energie et des Matériaux" (Orléans). The EELS analyses were performed at the "Laboratoire de Physique des Solides de l'Université d'Orsay" (France).

References-

Belin S. and Brosse E. (1992) Petrographical and geochemical study of a Kimmeridgian organic sequence (Yorskhire area, UK). *Rev. Inst. Fr. Pétr.*, **47**, 711-725.

Bertrand P., Lallier-Vergès E., Martinez L., Pradier B., Tremblay P., Huc A., Jouhannel R. and Tricart J.P. (1990) Examples of spatial relationships between organic matter and mineral groundmass in the microstructure of the organic-rich Dorset Formation rocks (Great Britain)". *Organic Geochem.,* **16**, 661-675.

Bertrand P. and Lallier-Vergès E. (1993) Past sedimentary organic matter accumulation and degradation controlled by productivity. *Nature,* **364**, 786-788.

Bertrand P., Lallier-Vergès E. and Boussafir M. (1994). Enhancement of both accumulation and anoxic degradation of organic carbon controlled by cyclic productivity: a model. *Org. Geochem.*, in press.

Bishop A.N. and Philp R.P. (1994) The potential for amorphous kerogen formation via adsorption of organic material at mineral surfaces. 207th American chemical Society 1994, National Meeting - March 13-18, San Diego.

Boussafir M., Lallier-Vergès E., Bertrand P. and Badaut-Trauth D. (1994a) Etude ultrastructurale de matières organiques micro-prélevées dans les roches de la "Kimmeridge Clay Formation" (Yorkshire, UK). *Bull. Centres Rech. Explor. Prod. Elf-Aquitaine*, 18, Publ. Spéc. 275-277.

Boussafir M., Lallier-Vergès E., Bertrand P. and Badaut-Trauth D. (1994b) Structure ultrafine de la matière organique des roches mères du kimméridgien du Yorkshire (UK). *Bull. de la Soc. Géol. de Fr.*, **165**(4), 355-363.

Derenne S., Largeau C., Casadevall E., Berkaloff C. and Rousseau B. (1991) Chemical evidence of kerogen formation in source rocks and oil shales via selective preservation of thin resistant outer walls of microalgae : Origin of ultralaminae. *Geochim. et Cosmochim. Acta.*, **55**, 1041-1050.

Derenne S., Le Berre F., Largeau C., Hatcher P., Connan J. and Raynaud J.F. (1993) Formation of ultralaminae in marine kerogens via selective preservation of thin resistant outer walls of microalgae. *Org. Geochem.*, in press.

Disko M. (1992) Transmission electron energy-loss spectrometry in materials science, DISKO M., AHN C.C. and FULTZ B (eds.). The Mineral, Metal and Material Society, Monograph series 2.

Henrichs S.M. and Reeburgh W.S. (1987) Anaerobic mineralization of marine sediment organic matter: rates and the role of anaerobic processes in the oceanic carbon economy. *Geomicrobiology Journal*, **5**, 191-237.

Herbin J.P., Müller C., Geyssant J.R., Mélières F. and Penn I.E. (1993) Variation of the Distribution of organic matter within a transgressive system tract: Kimmeridge Clay (Jurassic), England. *In*: A.A.P.G.1993 "Petroleum source rocks in a sequence stratigraphic framework". B. KATZ and L. PRATT (eds.). pp. 67-100.

Huc A.Y., Lallier-Vergès E., Bertrand P., Carpentier B. and Hollander D. J. (1992) Organic matter response to change of depositional environment in Kimmeridgian shales, DORSET, U.K. *In*: "Organic matter: productivity, accumulation and preservation of organic matter in recent and ancient sediments". JEAN WHELAN and JOHN FARRINGTON (eds). Columbia Univrsity Press. New York. pp. 469-486.

Jackson G.A. (1990) A model of the formation of marine algal flocs by physical coagulation processes. *Deep-Sea Research*, **37**(8), 1197-1211.

Jørgensen B.B. (1982) Mineralization of organic matter in the sea bed - the role of sulphate reduction. *Nature*, **296**, 643-645.

Lallier-Vergès E., Bertrand P., Boussafir M., Tribovillard N. and Desprairies A. (1992) Productivity as a major control of short-term organic cyclicity in the Kimmeridge rocks of Yorkshire (U.K.). Fourth International Conference on Palaeoceanography. Kiel, 21-25 septembre 1992.

Lallier-Vergès E., Boussafir M., Bertrand P. and Badaut-Trauth D. (1993) Selective preservation of various organic matter types as assessed by STEM studies on a cyclic productivity-controlled sedimentary series (Kimmeridge Clay Formation). *In* "Organic Geochemistry: Poster sessions from the 16th International Meeting on Organic Geochemistry, Stavanger, 1993". KJELL ØYGARD (ed.). Falch Hurtigtrykk, Oslo. pp. 384-386

Lallier-Vergès E., Bertrand P., Tribovillard N.P., Hayes J., Boussafir M., Zaback D.A. and Connan J. (1994) Productivity-induced sulfur enrichment of organic-rich sediments. 207th American Chemical Society 1994, National Meeting - March 13-18, San Diego.

Largeau C., Derenne S., Casadevall E., Berkaloff C., Corolleur M., Lugardon B., Raynaud J.F. and Connan J. (1990) Occurrence and origin of ultralaminar structures in "amorphous" kerogens from various source-rocks and oil-shales. *In:* Advances in organic geochemistry 1989. DURAND B. and BEHAR F. (eds.), pp. 889-896.

Lugardon B., Raynaud J.F. and Husson P. (1991) Données ultrastructurales sur la matière organique amorphe des kérogènes. *Palynoscience*, **1**, 69-88.

Pradier B. and Bertrand P. (1992) Etude à haute résolution d'un cycle de carbone organique des argiles du Kimméridgien du Yorkshire (GB): relation entre composition pétrographique du contenu organique observé in-situ, teneur en carbone organique et qualité pétroligène. *C.R. Acad. Sc. Paris*, **315**(2), 187-192.

Ramanampisoa L., Bertrand P., Disnar J.R., Lallier-Vergès E., Pradier B. and Tribovillard N.P. (1992) Etude à haute résolution d'un cycle de carbone organique des argiles du Kimméridgien du Yorkshire (GB): résultats préliminaires de géochimie et de pétrographie organique. *C.R. Acad. Sc. Paris*, **314**(2), 1493-1498.

Raynaud J.F., Lugardon B. and Lacrampe-Couloume G. (1988) Observation de membranes fossiles dans la matière organique "amorphe" de roches-mères de pétrole. *C. R. Acad. Sci. Paris*, **307**, 1703- 1709.

Tribovillard N.P., Desprairies A., Lallier-Vergès E., Bertrand P., Moureau N., Ramdani A. and Ramanampisoa L. (1994) Geochemical study of organic matter rich cycles from the Kimmeridge Clay Formation of Yorkshire (UK): productivity versus anoxia. *Palaeogeogr., Palaeoclim., Palaeoecol.*, **108**, 165-181.

Wefer G. (1989) Particles flux in the Oceans: effects of episodic production. *In* Report of the Dahlem Workshop on Productivity of the Oceans: Present and Past. BERGER W.H., SMETACEK V.S. and WEFER G. (eds.) Wiley, Chichester, 1989, pp. 139-154.

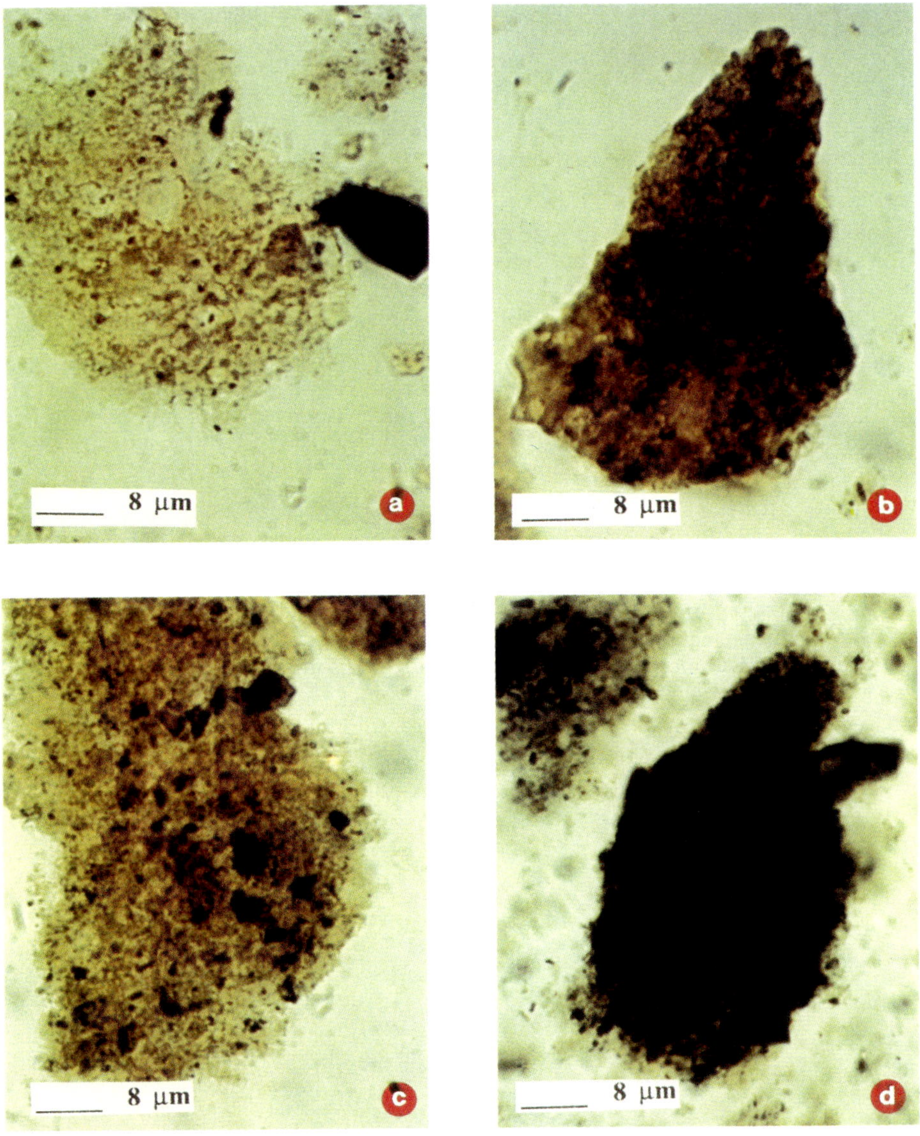

Plate 1 - The different types of organic matter observed by optical microscopy.(transmitted light); (a) yellow marine debris and black lignaceous debris; (b) orange amorphous organic matter; (c) brown amorphous organic matter; (d) black amorphous organic matter

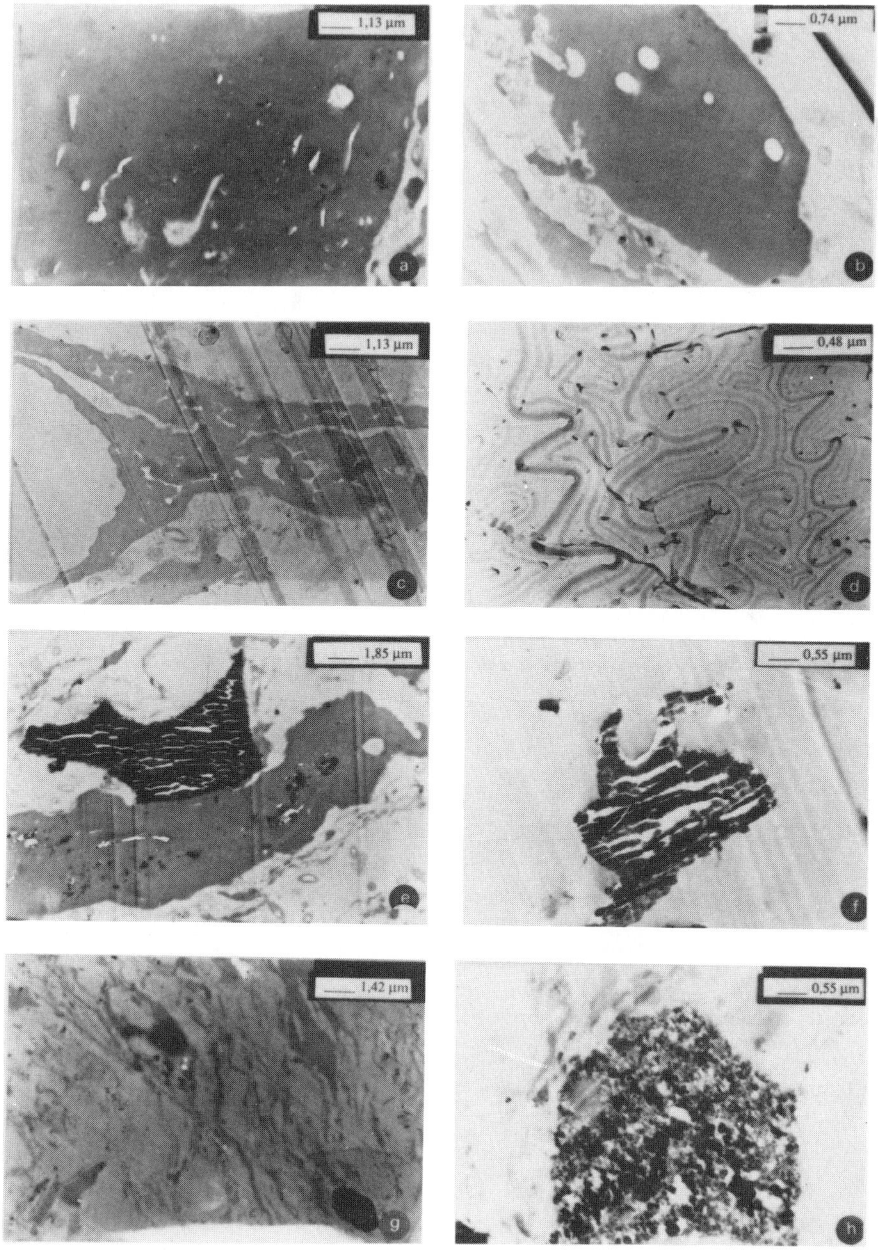

Plate 2 - Ultrafine structures of the different types of amorphous organic matter and from the bulk organic matter. (a, b) orange amorphous organic matter observed by TEM showing a nanoscopically amorphous and perfectly homogeneous ultrastructure; (c, d) brown amorphous organic matter observed by TEM exhibiting ultralaminar structures; (e) black amorphous organic matter observed by TEM; (f) lignaceous fragment in the nanoscopically amorphous organic material; (g) laminar diffuse amorphous organic matter occurring in the organo-mineral matrix; (h) granular organic material with aggregated spherules in which EDAX analyses indicate the occurrence of C, S and Fe.

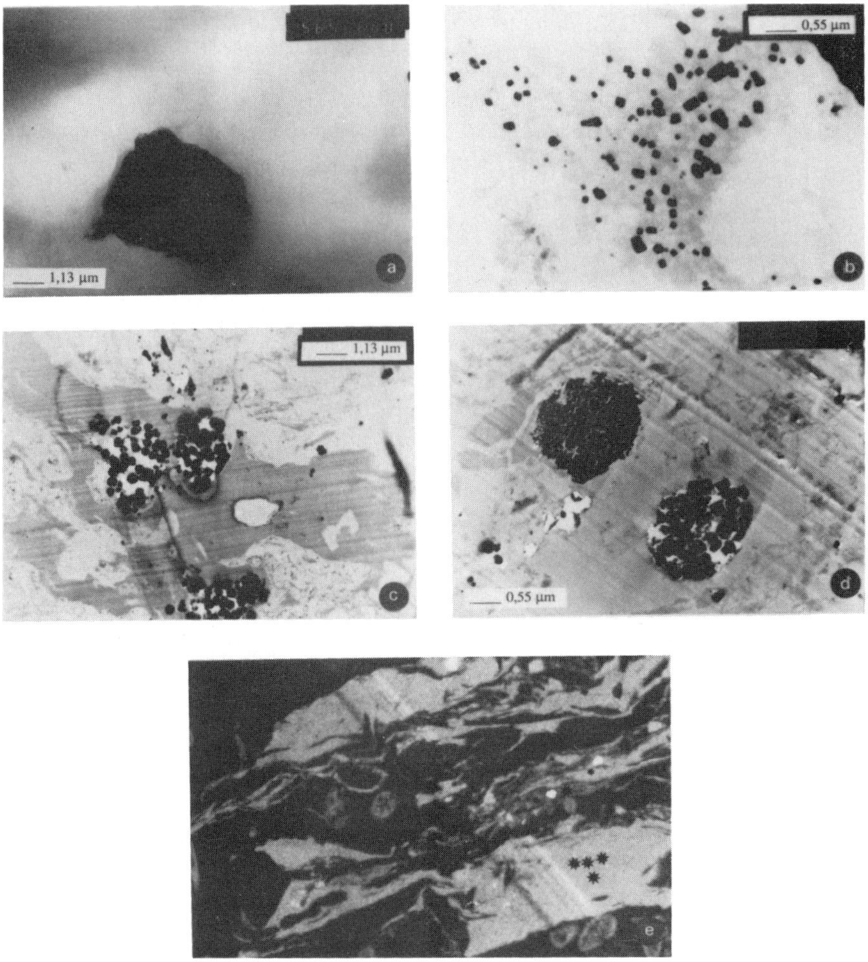

Plate 3 - Different occurrences of pyrite observed by TEM (a, b, c, d) - (a) automorphous isolated pyrite crystal; (b) assemblage of automorphous pyrite microcrystals; (c) nanoscopically amorphous and perfectly homogeneous organic matter with inclusions of pyrite framboids; (d) close association of a non-crystallised FeS spherule (left side of the micrograph) and framboidal pyrite. Location of EELS analyses in the nanoscopically amorphous organic matter (e) .

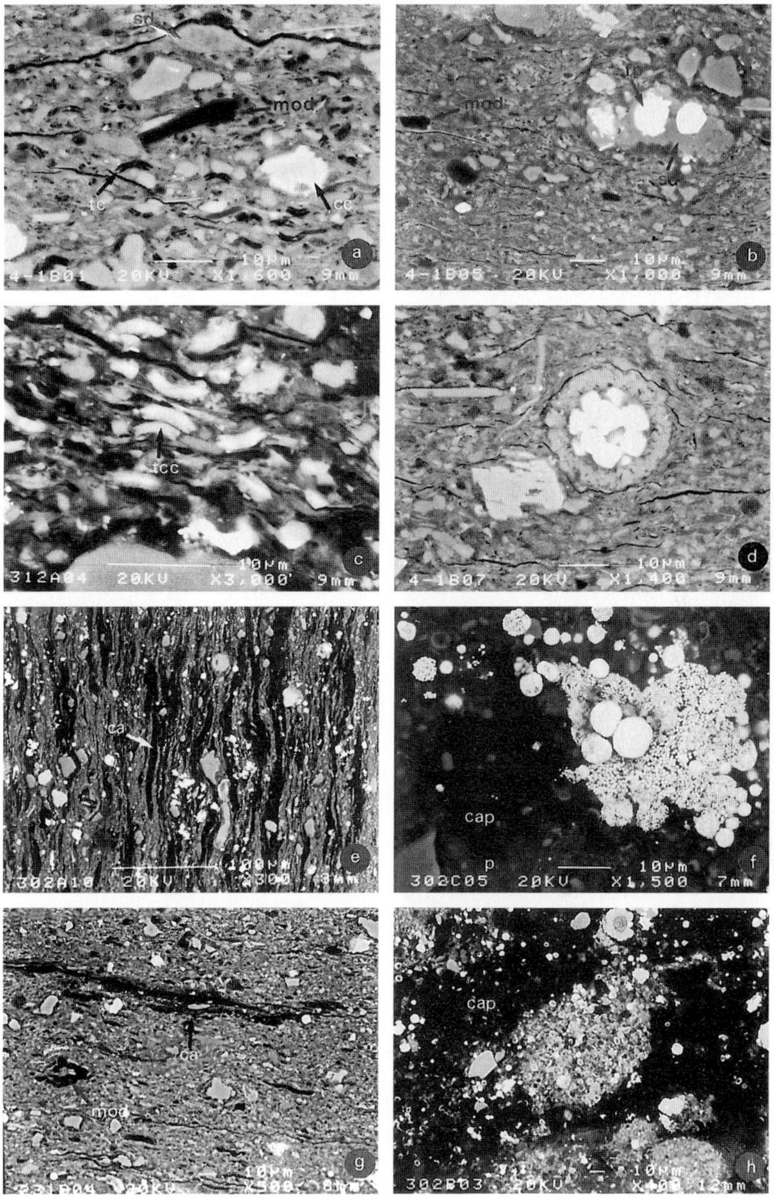

Plate 4 - SEM (back-scattered electron mode) micrographs performed on polished sections of the rock samples revealing the organic matter in black colour.
- (photos a, b, d) typical microfacies in low TOC samples (≈2%). The organic matter is dominated by terrestrial organic debris$_{(mod)}$. Micrograph (a) reveals coccolith prints filled with organic matter$_{(tc)}$. Pyrite crystals in microfossils (photo d); pyrite framboids$_{(fp)}$ occurring within diagenetic silica$_{(sd)}$ (photo b).
- (photos c, e, f, h) typical microfacies in high TOC samples (≈9%). Micrograph (c) reveals calcitic coccolith tests$_{(tcc)}$ and diffuse organic material (in black). Bituminite appears in cross-bedding sections (photo e) as elongated black bodies$_{(ca)}$ and in parallel-bedding sections as large-sized black bodies$_{(cap)}$ associated with pyrite framboids (photo f) and zooplankton pellets$_{(p)}$ composed of coccoliths (photo h).
- (photo g) typical microfacies in ≈ 4% TOC samples. Rare bituminite$_{(ca)}$ is present associated with terrestrial organic matter$_{(mod)}$.

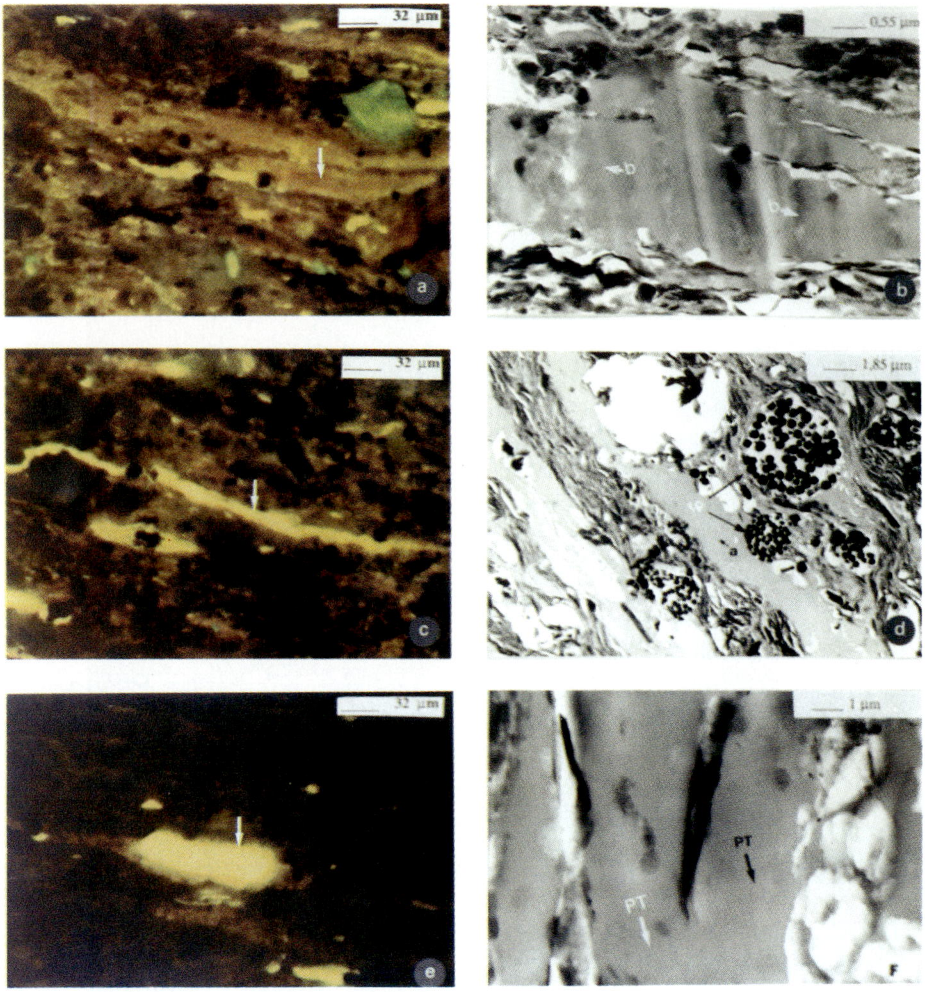

Plate 5 - Ultrafine structure of the three main macerals. (a) bituminite exhibiting a brown fluorescence under UV excitation; (b) bituminite$_{(b)}$ observed by TEM showing an internal amorphous ultrastructure; (c) alginite exhibiting a yellow fluorescence under UV excitation; (d) alginite$_{(a)}$ observed by TEM showing an internal amorphous ultrastructure. Pyrite framboids$_{(fp)}$ are also visible on the micrograph; (e) *Tasmanacea* exhibiting a high yellow fluorescence under UV excitation; (f) *Tasmanacea* observed by TEM revealing thick amorphous walls$_{(PT)}$.

Study of qualitative and quantitative variations in kerogen chemical structure along a microcycle: Correlations with ultrastructural features

François Gelin[1], *Mohammed Boussafir*[2], *Sylvie Derenne*[1], *Claude Largeau* [1] *and Philippe Bertrand*[2]

1) Ecole Nationale Supérieure de Chimie de Paris, 11 rue Pierre et Marie Curie, F-75231 Paris cedex 5
2) Université d'Orléans, URA 724 du CNRS, Dépt. des Sciences de la Terre, F-45067 Orléans cedex

Key-words- *Kimmeridge Clay Formation, pyrolyses, kerogen chemical structure, source organisms, TOC and HI cyclic variations.*

Abstract- Kerogens corresponding to typical points of a Kimmeridge Clay microcycle (beginning, top, end) were examined by Curie point and "off-line" pyrolyses. Pyrolysis products (hydrocarbons, thiophenes, medium polarity compounds, phenols, fatty acids) were identified and their relative abundances determined. These pyrolysis experiments provided information on (i) the chemical features, source organisms and mode of formation of the three types of organic matter identified by microscopy, (ii) the causes of the large variations in the abundances of the above types occurring along a microcycle and of the resulting changes in kerogen quantity (TOC) and quality (HI) and (iii) the origin of microcycles. Selective Preservation played a major role in kerogen formation at the beginning and the end of microcycles whereas lipid "vulcanisation" was the predominant process at the top. These chemical results confirmed that TOC and HI microcycles in Kimmeridge Clay reflect variations in primary productivity that are strongly amplified by associated changes in sulphate reduction intensity.

Introduction-

The Kimmeridge Clay formation consists of organic-rich sediments which were deposited in cyclic short-time periods, termed microcycles, with respect to their organic matter (OM) (Herbin *et al.*, 1991). As described in the related studies of the project, light microscopy allowed the identification of three different types of OM which are always present along the microcycle, but with highly variable relative abundances (Ramanampisoa *et al.*, 1992; Lallier-Vergès *et al.*, 1993). Parallel examinations of the three OM by light microscopy and transmission electron microscopy or TEM (Boussafir *et al.*, 1994) lead to the following findings: the "brown OM" is mainly composed, as revealed by TEM, of ultralaminae-like structures that could not be detected by light microscopy and is dominantly present at the beginning and the end of the microcycle. The "orange OM", nanoscopically amorphous, dominates in the middle of the microcycle, *i.e.* when the total organic carbon content (TOC) and the hydrogen index (HI) show maximal values. The "black OM", which is mainly composed of minute plant-derived debris not identified by light microscopy associated with nanoscopically amorphous OM particles, always exhibits a fairly constant and weak relative abundance; its contribution is especially low at the top of the microcycle.

The framework of this study was the analysis of three samples belonging to the same microcycle and the major aims were (i) to define the chemical characteristics of these different types of organic matter identified by microscopy, (ii) to establish relationships between such chemical features and both source organisms and depositional conditions and (iii) to understand better the origin of the changes occurring in kerogen quantity and quality during a microcycle. To this end, pyrolytic studies of samples corresponding to typical points on a microcycle were carried out and pyrolysate constituents were identified by gas chromatography-mass spectrometry (GC-MS).

Experimental-

Samples-

As indicated in Table 1, three samples (termed A, B and C) were selected along the same microcycle with respect to TOC and HI. Samples A, B and C correspond to the beginning, the top and the end of the cycle, respectively. Kerogens were isolated after bitumen extraction and the usual HF/HCl treatment.

	Depth (m)	TOC(%)	HI[a]
A	128.233	2.82	495
B	128.755	9.51	582
C	129.093	1.74	243

[a] mg of hydrocarbons/g of organic C.

Table 1 - Location and bulk geochemical data.

Pyrolysis-

"Off-line" pyrolysis was performed according to Largeau *et al.* (1986). Briefly, kerogens are first heated at 300°C for 20 min, after extraction with $CHCl_3$/MeOH (2/1), the insoluble residue is pyrolysed at 400°C during 2 hr. Helium is used as carrier gas. The released products are trapped in $CHCl_3$ at -5°C and fractionated as depicted in fig. 1. The various fractions are analysed by GC-MS.

Curie point pyrolysis-gas chromatography-mass spectrometry (Py-GC-MS) was also performed on samples A and B using a Curie point pyrolyser (FOM3-LX) and a ferromagnetic wire with a Curie temperature of 610°C.

Results-

The Py-GC traces of samples A and B (fig.2) show a highly dominant homologous series of *n*-alk-1-ene/*n*-alkane doublets ranging from C_7 to C_{30} and there is no significant difference in the relative intensity and distribution of this series for the two pyrolysates. The formation of such compounds upon pyrolysis is known to reflect the presence of long polymethylenic chains. The Py-GC traces also show a high abundance of compounds eluting between the alkene/alkane doublets. As indicated in fig. 2, some of these products were identified as series of alkylated thiophenes, alkylated phenols, alkylated benzenes and linear/branched, (un)saturated hydrocarbons. The major difference that can be noted between the two pyrolysates is the higher relative abundance in thiophenic compounds for sample B. This difference can be assessed by the thiophene ratio (ThR*). ThR is known to provide a convenient way for a rapid estimation of total organic sulfur contribution in kerogens (Eglinton *et al.*, 1990). Indeed a substantially higher ThR (0.22 against 0.12) is noted for sample B when compared to A, thus indicating a higher participation of sulfur-containing moieties to kerogen in the middle of the microcycle.

Due to the high complexity of the flash pyrolysates, a more detailed analysis of the pyrolysis products can only be achieved by "off-line" pyrolysis and identification of the constituents of the various fractions resulting from the separation process (fig. 1). The balance of the 400 °C pyrolyses and the yields of the pyrolysates fractions are reported in Table 2** . Table 3 shows the different homologous series identified in the hexane sub-fractions of the pyrolysates of samples A and B. In addition to the major series of *n*-alkanes already discussed, the first subfraction comprises various series of alkanes, either branched or containing a C_6 or C_5 ring associated with the *n*-alkyl chain. The presence, in the case of sample A, of two series corresponding to odd carbon numbered 3,7-dimethylalkanes and even carbon numbered 3-methylalkanes is noteworthy. These two series were observed, with similar distributions and relative abundances, in the pyrolysates of the non-hydrolysable, resistant macromolecules located in several bacteria and termed bacterans (Le Berre *et al.*, 1991; Flaviano *et al.*, 1994). The second and third sub-fractions, in addition to *n*-alkenes and *n*-alkadienes, comprise several series of thiophenic and benzenic compounds.

---

* ThR = [2,3-dimethylthiophene]/([non-1-ene]+[1,2-dimethylbenzene]) is calculated from the relative abundance of these compounds in the flash pyrolysate.

** As indicated in this Table, GC/SM analyses were only carried out, for kerogen C, on the crude pyrolysate. Such analyses revealed bulk features similar to those observed for sample A (*e.g.* a similar production of *n*-alkylnitriles; see also Fig. 3).

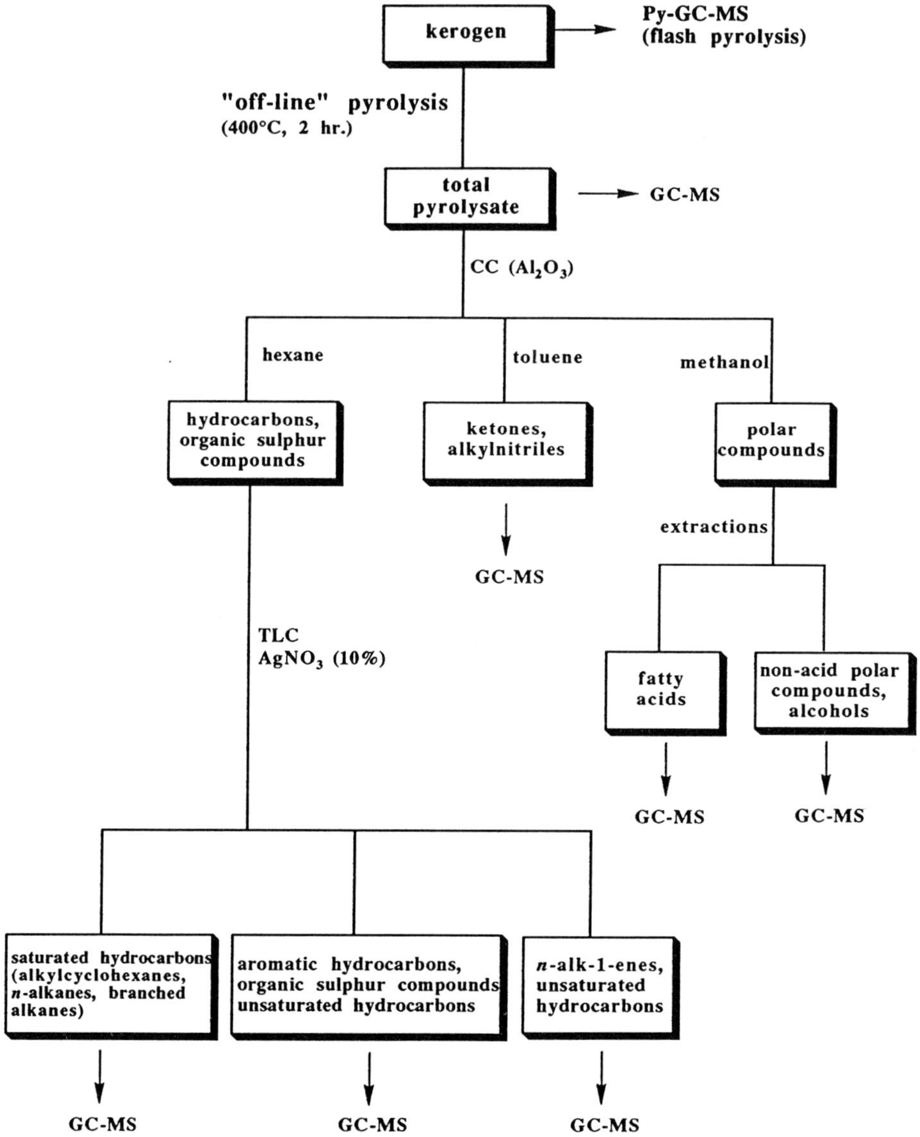

Fig. 1. Analytical flow diagram for "off-line" pyrolysis.

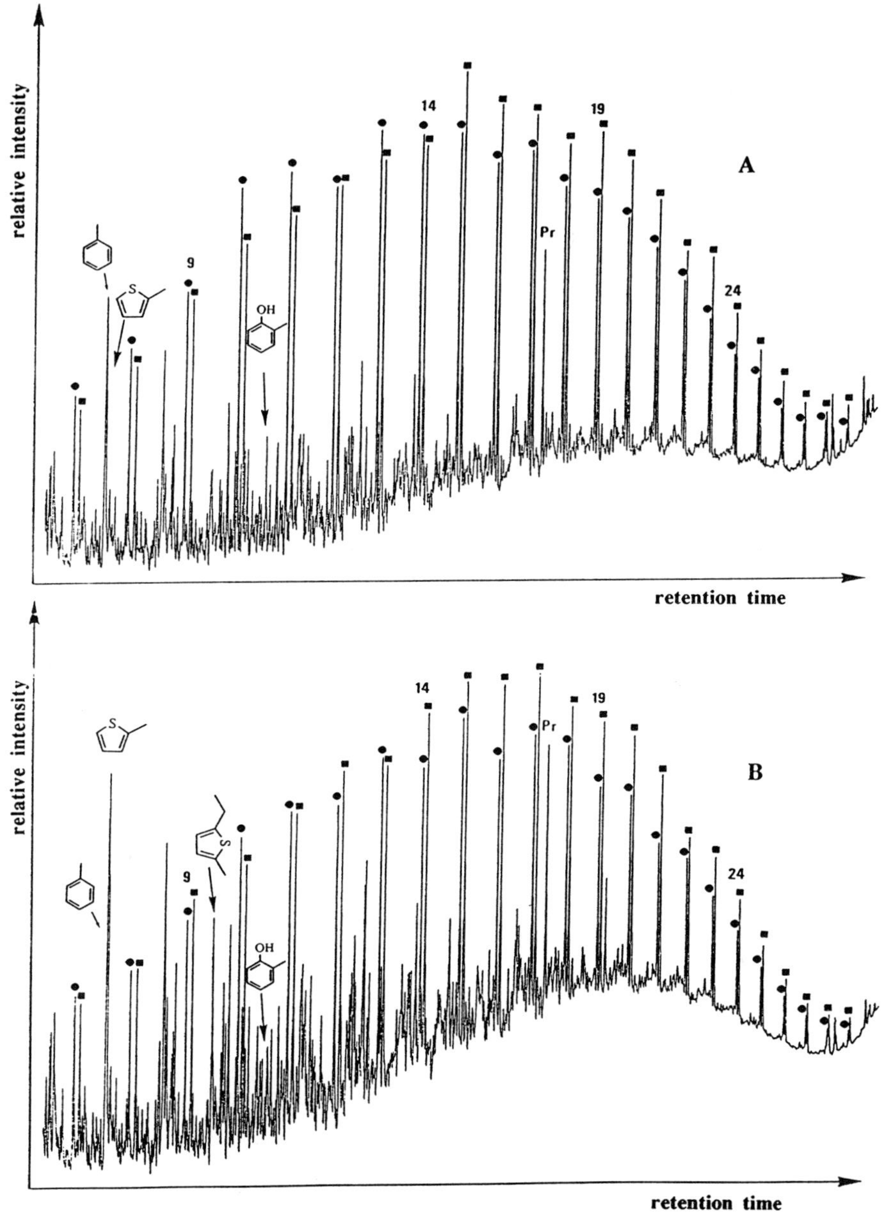

Fig. 2. Total ion current (TIC) traces of the flash pyrolysates of samples A and B. Filled circles and filled squares indicate the homologous series of *n*-alk-1-enes and *n*-alkanes, respectively. Numbers indicate their chain length. The structure of some major compounds is indicated and Pr designates prist-1-ene.

	Balance of pyrolysis[b]				Crude pyrolysate composition[c]		
	Weight loss	Trapped	Extracted[d]	Residue	Hexane eluted	Toluene eluted	Methanol eluted
A	28.0	12.8	0.9	54.9	37	19	44
B	55.2	37.8	0.8	30.3	32	27	41
C	20.8	5.2	1.5	60.5	n.d.[e]	n.d.[e]	n.d.[e]

[a] High volatility products are not recovered by this method; their abundance (%) can be calculated from weight loss (%) - trapped (%).
n.d.: not determined
[b] As % of the initial kerogen.
[c] Relative abundances of the three isolated fractions as % of the total trapped compounds.
[d] High molecular weight soluble pyrolysis products not swept away by the He flow.
[e] No CC fractionation was carried out in that case due to the low amount of pyrolysate obtained; GC/MS analysis was directly performed on the crude pyrolysate.

Table 2 - Pyrolysis at 400 °C[a].

One of the major differences between the samples is the much higher relative abundance of the thiophenic compounds for sample B (corresponding to the highest TOC and HI). For example, the alkylated benzothiophenes are very abundant in the pyrolysate of sample B while only present as traces in the pyrolysate of sample A. This observation, which confirmed the previously discussed flash pyrolysis results, can be illustrated by comparison of the two mass chromatograms of *m/z* 111 of the pyrolysates of kerogens A, B and C (fig. 3), highlighting the series of 2-alkyl,5-methylthiophenes and *n*-alk-1-enes. This figure clearly shows that the former are more abundant with regard to the alkenes in the case of sample B. This difference in the relative abundances of the thiophenic compounds reveals that "natural vulcanisation" (sulfur incorporation into lipids during early diagenesis) has probably contributed much more to the formation of the kerogen from the middle of the microcycle. Owing to such a "vulcanisation" process, some lipids are rapidly incorporated into the insoluble high molecular weight fraction and thus escape biodegradation in participating to the much more resistant macromolecular system so formed. This process therefore provides a very high "acquired" resistance to lipids that, otherwise, would be heavily degraded during deposition.

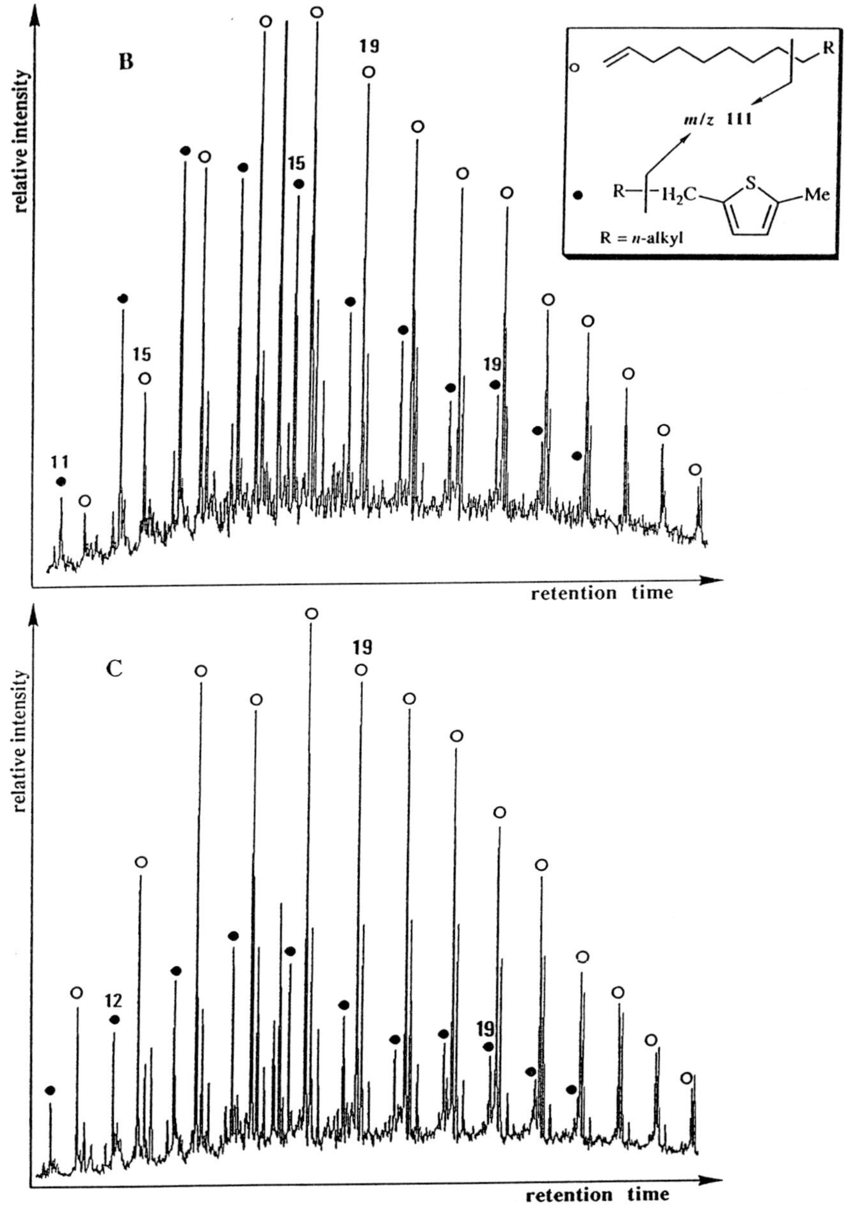

Fig. 3. Mass chromatograms of *m/z* 111 revealing the homologous series of 2-alkyl, 5-methylthiophenes (filled circles) and *n*-alk-1-enes (empty circles) in the total "off-line"pyrolysates of samples B and C. Numbers indicate the length of the alkyl chains. Similar mass chromatograms were obtained for A and C.

	A		B	
	constituents	r.a.	constituents	r.a.
n-alkanes	C_{14}-C_{31} (C_{17})	1	C_{12}-C_{30} (C_{15})	1
isoprenoid alkanes	C_{15}-C_{20} (C_{18})	0.16	C_{15}-C_{20} (C_{18})	0.23
3,7-dimethylalkanes	C_{17}-C_{31} (C_{25})	0.08	n.d.	—
3,5-dimethylalkanes	C_{15}-C_{31} (C_{21})	0.05	n.d.	—
3-methylalkanes	C_{14}-C_{30} (C_{20})	0.10	C_{13}-C_{22} (C_{15})	0.04
n-alkylcyclopentanes	C_{14}-C_{26} (C_{20})	0.27	n.d.	—
n-alkylcyclohexanes	C_{14}-C_{25} (C_{19})	0.05	C_{12}-C_{23} (C_{14})	0.13
n-alkenes (E)	C_{14}-C_{26} (C_{17})	1	C_{13}-C_{25} (C_{15})	1
n-alkadienes[b]	C_{14}-C_{24} (C_{17})	0.52	C_{12}-C_{22} (C_{15})	0.62
n-alkylbenzenes	C_{14}-C_{22} (C_{16})	0.16	C_{12}-C_{18} (C_{14})	0.35
n-alkyltoluenes	C_{14}-C_{23} (C_{16})	0.18	C_{12}-C_{21} (C_{15})	0.28
n-alkyldimethylbenzenes	C_{14}-C_{20} (C_{16})	0.16	C_{12}-C_{18} (C_{14})	0.33
2-*n*-alkylthiophenes	C_{12}-C_{24} (C_{14})	0.15	C_{10}-C_{24} (C_{12})	0.65
2-*n*-alkyl,5-methylthiophenes	C_{12}-C_{24} (C_{14})	0.33	C_{10}-C_{20} (C_{12})	0.86
n-alkyldimethylthiophenes	C_{12}-C_{24} (C_{15})	0.06	C_{10}-C_{20} (C_{12})	0.28
n-alk-1-enes	C_{14}-C_{28} (C_{18})	1	C_{12}-C_{28} (C_{15})	1
n-alkenes (Z)	C_{14}-C_{25} (C_{18})	0.09	n.d.	—
n-alkadienes[b]	C_{14}-C_{22} (C_{18})	0.08	C_{12}-C_{18} (C_{14})	0.11
alkylbenzothiophenes	C_8-C_{12}	trace	C_8-C_{12} (C_{11})	0.40
alkylnaphthalenes	C_{10}-C_{14}	trace	C_{10}-C_{14} (C_{12})	0.15

[a] r.a.: relative abundance of the homologous series calculated with respect to the three predominant series of the three sub-fractions; the bracketed values correspond to the maximum of each series. n.d.: not detected.
[b] The double bond positions could not be determined. The alkadiene series of the second and third sub-fractions exhibit different retention times and do not correspond to α,ω-alkadienes.

Table 3 - Nature and relative abundance of the homologous series of compounds identified in the hexane sub-fractions of the 400°C pyrolysates of samples A and B[a].

Table 4 shows the different series of compounds identified in the toluene fraction of the pyrolysates of samples A and B. In both pyrolysates, the fractions are dominated by a series

Lecture Notes in Earth Sciences, vol. 57. Erratum p. 151.

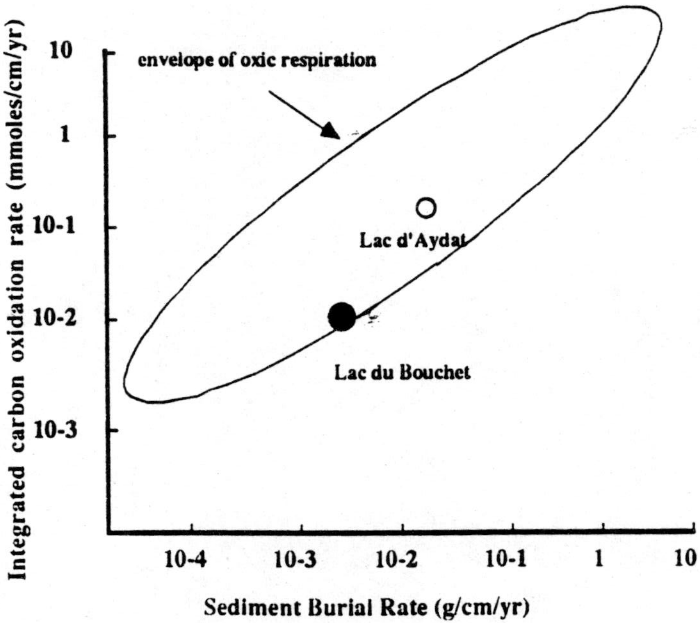

Figure 6. Diagram of the integrated carbon oxidation rate versus the sedimentation rate for sediments from the Lac du Bouchet (solid circle) and the Lac d'Aydat (hollow circle). The envelope of oxic respiration is shown after Canfield (1989).

Lecture Notes in Earth Sciences, Vol. 57
Lallier-Vergès, Tribovillard, Bertrand,
Organic Matter Accumulation

of ketones identified as *n*-alkan-2-ones. These ketones, commonly observed in pyrolysates of various kerogens (Van de Meent *et al.*, 1980) are probably derived from the pyrolysis of macromolecules with cross-linking ether bonds (Gelin *et al.*, 1993). The significance of the presence of series of indoles and quinolines, although abundant, has not been determined so far.

	A		**B**	
	constituents	r.a.	constituents	r.a.
n-alkylmethylketones	C_{14}-C_{31} (C_{17})	1	C_{10}-C_{23} (C_{13})	1
branched alkylmethylketones	n.d.	—	C_{11}-C_{21} (C_{13})	0.3
n-alkylethylketones	n.d.	—	C_{10}-C_{25} (C_{13})	0.2
n-alkylnitriles	C_{11}-C_{21} (C_{14})	0.8	n.d.	—
ethylalkanoates	n.d.	—	C_{12}-C_{22} (C_{18})	traces
alkylated indoles	C_9-C_{12}	0.9	C_9-C_{11}	0.1
alkylated quinolines	C_{10}-C_{12}	0.15	C_{10}-C_{13}	0.3

[a] r.a.: relative abundance of the homologous series calculated with respect to the predominant series. n.d.: not detected. The bracketed values correspond to the maximum of each series.

Table 4 - Nature and relative abundance of the homologous series of compounds identified in the toluene fraction of the 400 °C pyrolysates of samples A and B [a].

However, the main information obtained from this fraction is the significant presence of a series of *n*-alkylnitriles in the pyrolysate of sample A while it is not detected in the case of sample B. *n*-Alkylnitriles have only been observed so far in the pyrolysates of kerogens exhibiting ultralaminae when observed by TEM (Largeau *et al.*, 1990; Derenne *et al.*, 1991, 1992a). These studies revealed that, based on nitrile specific distributions, it is possible to determine the marine or lacustrine origin of the tested kerogen (Derenne *et al.*, 1992b). Moreover, it was found that the algaenan*- composed thin outer walls isolated from several extant freshwater and marine microalgae are morphologically very similar to the above ultralaminae and release the same *n*-alkylnitriles upon pyrolysis (Derenne *et al.*, 1991, 1992a). The presence of these nitriles in pyrolysates is therefore considered as an indicator for the contribution of such thin resistant outer walls of microalgae to kerogen formation *via*

* The general term algaenan is used for the non-hydrolysable highly aliphatic, insoluble biomacromolecules present in microalgal cell walls (Tegelaar *et al.*, 1989, de Leeuw and Largeau, 1993).

the selective preservation pathway. This pathway is based on the occurrence of insoluble and non-hydrolysable macromolecules (algaenans for algae) in the source organisms (Tegelaar *et al.*, 1989; Largeau and Derenne, 1993). While the other cell constituents are rapidly degraded during the first steps of sedimentation, these intrinsically resistant polymers are selectively preserved in kerogens. This mechanism is thus sharply different from the classical Degradation-Recondensation pathway of kerogen formation (Tissot and Welte, 1984) which was the only mechanism considered till the last few years.

The same series of alkylated phenols are present in the pyrolysates of the studied kerogens, but their relative abundances vary significantly depending on the considered sample; thus phenol derivatives are 1.5 to 2 times more abundant in the pyrolysates of samples A and C than in the case of sample B. Alkylated phenols in pyrolysates are generally relevant for the contribution of terrestrial OM to kerogens. Indeed, phenols are the major pyrolysis products of diagenetically modified lignins (Larter, 1984; Saiz-Jimenez and de Leeuw, 1986). The summed mass chromatograms of *m/z* 107+108 shown in fig. 4 reveal the distribution of the two main series for the pyrolysate of sample C. These phenols appear to be dominated by mono- and dimethylphenols that are known as typical pyrolysis products of fossil lignins.

The acid fractions isolated from the three pyrolysates show similar relative abundances and distributions (fig. 5). The nature and the distribution of these acids can give important information about OM preservation during sedimentation and burial (Flaviano *et al.*, 1994). For each sample, the predominant saturated normal fatty acids range from C_{14} to C_{24} and their distribution reveals a strong even over odd carbon number predominance (CPI, calculated in the C_{14}-C_{20} range, of 0.12, 0.10 and 0.15 for samples A,B and C, respectively). This predominance is characteristic of biological systems. The presence in relatively high amounts of monounsaturated C_{16} and C_{18} fatty acids should also be noted. Unsaturated acids, when in free lipids, are known to resist diagenetic degradation poorly. Taken together, the distribution of the saturated acids and the occurrence of unsaturated acids in the pyrolysates of kerogens A, B and C indicate a good preservation of the source materials. Such a feature can reflect either the presence of the corresponding esters in selectively preserved resistant macromolecules like algaenans or early "vulcanisation" of ester-containing lipids.

Discussion and conclusions-

Table 5 summarises the main information obtained from the study of the pyrolysis products of samples A-C and correlates them, when possible, with OM morphological features observed by light and electron microscopy:

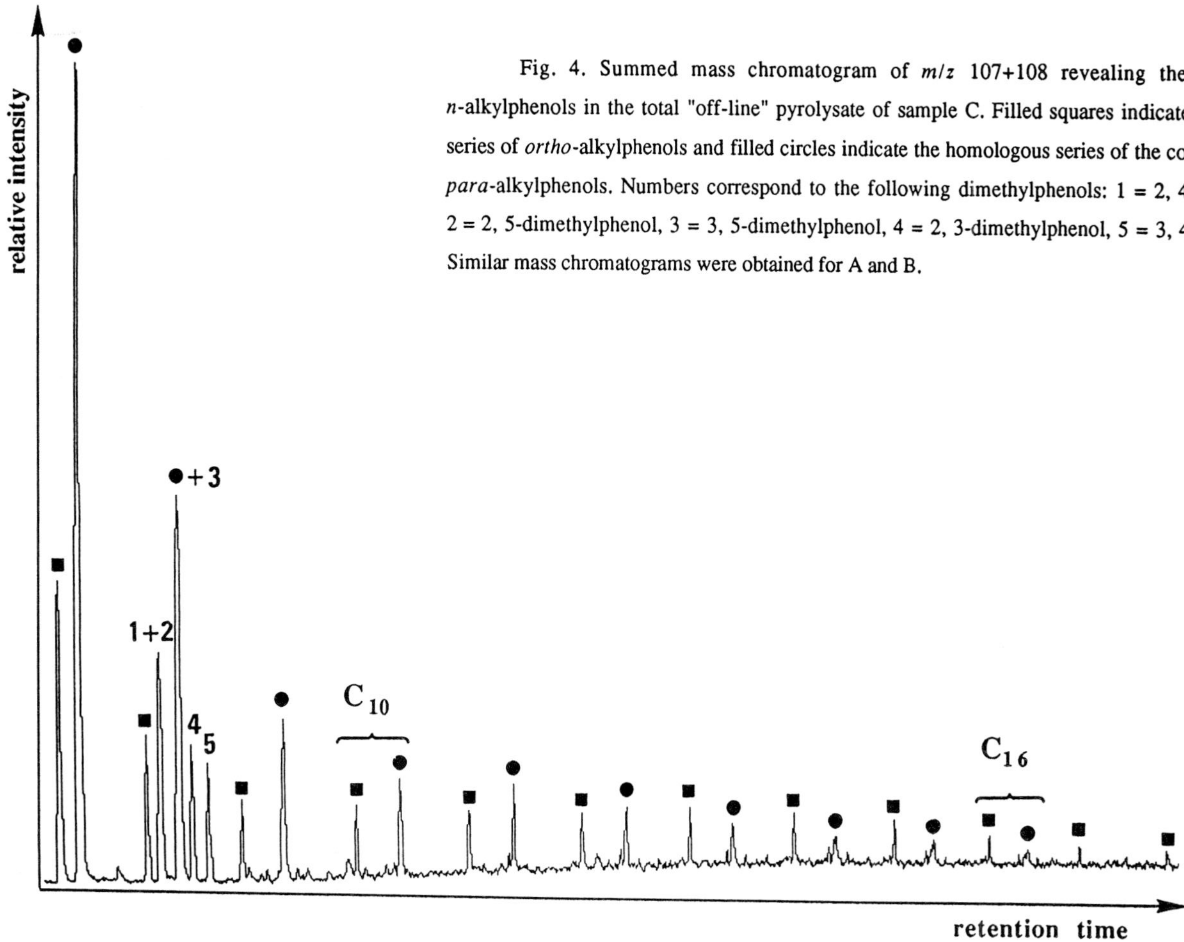

Fig. 4. Summed mass chromatogram of m/z 107+108 revealing the distribution of n-alkylphenols in the total "off-line" pyrolysate of sample C. Filled squares indicate the homologous series of *ortho*-alkylphenols and filled circles indicate the homologous series of the coeluting *meta*- and *para*-alkylphenols. Numbers correspond to the following dimethylphenols: 1 = 2, 4-dimethylphenol, 2 = 2, 5-dimethylphenol, 3 = 3, 5-dimethylphenol, 4 = 2, 3-dimethylphenol, 5 = 3, 4-dimethylphenol. Similar mass chromatograms were obtained for A and B.

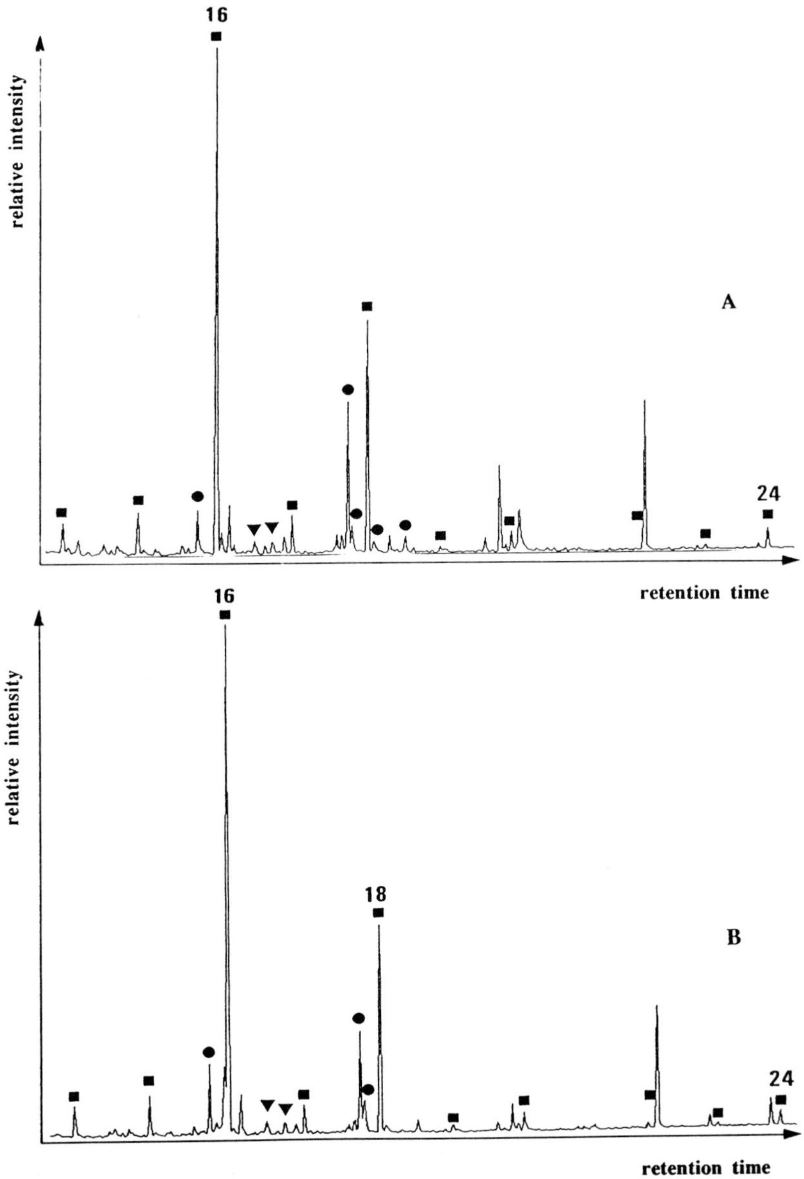

Fig. 5. TIC traces of the acid fractions from "off-line" pyrolysates of samples A and B. Filled squares and circles indicate saturated and unsaturated fatty acids, respectively; filled triangles indicate branched unsaturated acids. A similar trace was obtained for sample C.

- The organic sulfur compounds reflect a "vulcanisation" process due to the fast reaction, at the early diagenesis stage, of polysulphides and hydrogen sulphide with various free lipids. This phenomenon increased the preservation potential of the OM in creating a more resistant macromolecular system. TOC is thus increased in the corresponding sediments and the so-formed kerogen is amorphous. Consequently, the orange amorphous OM, dominantly present when TOC is maximum, must originate from lipid "vulcanisation", as confirmed by thiophenic compound abundance.

Pyrolysis products	Process of OM formation	*Optical OM type* TEM OM type
Organic sulphur compounds	Lipid "vulcanisation"	*Orange amorphous OM* Nanoscopically amorphous OM
Alkylnitriles	Algaenan selective preservation	*Brown amorphous OM* Ultralaminae
Alkylphenols	Lignin preservation	*Black amorphous OM* Minute lignaceous debris
Fatty acids	Selective preservation or "vulcanisation"	*
Branched alkanes	Bacteran selective preservation	*Brown & black amorphous OM* Diffuse nanoscopically amorphous OM (matrix)

* Fatty acids cannot be related to a given OM type; they can be associated with any fraction except the lignin-derived debris in the black OM.

Table 5 - Relationships between pyrolysis products, formation process and organic matter type in Kimmeridge Clay kerogen.

- *n*-Alkylnitriles are specific pyrolysis products of ultralaminae. Such compounds occur in substantial amounts in the pyrolysates of kerogens A and C, *i.e.* in the case of the samples dominated by brown OM. These chemical features, added to TEM observations, demonstrate that this brown OM originates from the selective preservation of resistant outer walls of microalgae.

- Alkylated phenol production upon pyrolysis reveals the presence of lignin-derived material. Such phenols appear to be relatively more abundant from samples A and C when compared to B. Previous observations on Kimmeridge Clay kerogens indicated that the black organic matter comprises minute lignaceous debris. Moreover, the contribution of this type of matter was shown to be low all along the microcycle and to be especially weak at the top, *i.e.* for samples like B. The above features, added to the present chemical observations, indicate a terrestrial origin for this fraction of the black organic matter of the Kimmeridge Clay and an important role of lignin-derived products in its formation.

- Fatty acid distributions indicate a relatively good preservation for the three kerogens that is due either to the incorporation of these acids via "vulcanisation" or to their occurrence as esters within the macromolecular network of the preserved algaenans.

- A major relative contribution of the selective preservation process in kerogen formation at the beginning and the end of the microcycle is also supported by the presence of some branched alkanes with specific structures and distributions in pyrolysates, that probably indicate a contribution of selectively preserved bacterans. TEM observations on bacterans isolated from extant species revealed an amorphous nature (Flaviano *et al.*, 1994). Accordingly, the selective preservation of such compounds can only be defined at the molecular level. Since specific pyrolysis products of such resistant biomacromolecules are not observed from kerogen B, the presence of selectively preserved bacterans in the orange organic matter is unlikely. The bulk of this fossil material must therefore be located in the brown and/or black organic matter and corresponds to the amorphous matrix occurring in the above two types along with ultralaminae and ligneous debris, respectively.

Origin of microcycles-

Previous studies related to this project showed that the parallel increase in the abundance (TOC) and the quality (HI) of the fossilised OM, observed in the middle of the microcycle, can be related to an increasing primary productivity (Bertrand and Lallier-Vergès, 1993) and that the intensity of sulphate reduction was a major parameter in TOC control. The present chemical observations fully confirm the crucial role of such processes.

As discussed in other papers of this volume (Boussafir *et al.*; Lallier-Vergès *et al.*), an increase in primary productivity is associated with a higher sulphate-reducing activity, due to a more efficient exportation of metabolisable organic matter to the anoxic zone. Such a situation can trigger a strong "vulcanisation" of lipids and hence reduce the biodegradation of the latter compounds, owing to their incorporation into macromolecular structures. The top of the TOC microcycle thus corresponds to a marked predominance of the orange amorphous organic matter formed, as shown here, *via* lipid "vulcanisation". On the contrary, a low primary productivity will be associated with a low activity of sulphate reducers and an extensive degradation of the metabolisable products will occur. As a result, a relatively small amount of organic matter will be finally buried in sediments and the low TOC thus obtained will mainly correspond to resistant macromolecular compounds. Kimmeridge Clay kerogens at the end and the beginning of the microcycle are thus chiefly composed of brown organic material derived from algaenan selective preservation and of black material derived from lignins. In addition, selectively preserved bacterans shall contribute to the amorphous matrix of the black and/or brown OM.

The above features also account for HI variations. High HI values are expected for the orange organic matter, since the lipids that are easily "vulcanised" are known to be highly aliphatic and based on long alkyl chains. The TOC - rich samples at the top of the microcycle are thus characterised by strong HI because they are dominated by this orange matter. Algaenans are aliphatic macromolecules based on a network of long alkyl chains. Accordingly, the important contribution of brown, ultralaminae - composed, organic matter at the end and beginning of the microcycle should favour high HI values. However, as already stressed, the brown material is then associated with substantial amounts of lignin - derived black organic matter and it is well documented that lignin - derived compounds are characterised by extremely low HI. Hence the relatively weak hydrogen indexes observed for the kerogen samples corresponding to the beginning and the end of the microcycle.

The occurrence of microcycles in the Kimmeridge Clay Formation reflects variations in primary productivity. However, the latter factor did not exert a single and direct control on TOC and HI. In fact, the intensity of sulphate reduction appears as a major parameter that strongly amplified the consequences of these primary productivity variations. The quantity of kerogen in the Kimmeridge Clay (TOC) is determined by the contributions of both macromolecular compounds with a high intrinsic resistance to diagenetic degradations, like algaenans and lignins, and lipidic components that acquired a resistant nature *via* "vulcanisation". Kerogen quality (HI) reflects the balance between these different contributions ; high HI values are promoted by large levels of "vulcanised" lipids and/or selectively preserved algaenans.

Acknowledgments- The authors are grateful to Dr. Jan W. de Leeuw for having allowed FG to perform the Curie point pyrolysis analyses.

References-

Bertrand P. and Lallier-Vergès E. (1993) Past sedimentary organic matter accumulation and degradation controlled by productivity. *Nature,* **364**, 786-788.

Boussafir M., Lallier-Vergès E., Bertrand P. and Badaut-Trauth D. (1994) Structure ultrafine de la matière organique des roches mères du kimméridgien du Yorkshire (UK). *Bull. Soc. Géol. Fr.*, **165**(4), 355-363.

Derenne S., Largeau C., Casadevall E., Berkaloff C. and Rousseau B. (1991) Chemical evidence of kerogen formation in source rocks and oil shales *via* selective preservation of thin resistant outer walls of microalgae : origin of ultralaminae. *Geochim. Cosmochim. Acta*, **55**, 1041-1050.

Derenne S., Le Berre F., Largeau C., Hatcher P., Connan J. and Raynaud J.F. (1992a) Formation of ultralaminae in marine kerogens *via* selective preservation of thin resistant outer walls of microalgae. *Org. Geochem.*, **19**, 345-350.

Derenne S., Largeau C. and Hatcher P. G. (1992b) Structure of *chlorella fusca* algaenan: relationships with ultralaminae in lacustrine Kerogens; species- and environment-dependent variations in the composition of fossil ultralaminae.*Org. Geochem.*, **18**(4), 417-422.

Eglinton T. I., Sinninghe Damsté J. S., Kohnen M. E. L. and de Leeuw J. W. (1990) Rapid estimation of the organic sulphur content of kerogens, coals and asphaltenes by pyrolysis-gas chromatography. *Fuel*, **69**, 1394-1404.

Flaviano C., Le Berre F., Derenne S., Largeau C. and Connan J. (1994) First indications of the formation of kerogen amorphous fractions by Selective Preservation. Role of non-hydrolysable macromolecular constituents of Eubacterial cell walls *Org. Geochem.,* in press.

Gelin F., Gatellier J.-P., Sinninghe Damsté J. S., Derenne S., Largeau C., Metzger P. and de Leeuw J. W. (1993) Mechanisms of flash pyrolysis of ether lipids isolated from the green microalga *Botryococcus braunii* Race A. *J. Anal. Appl. Pyrolysis,* **27**, 155.

Herbin J. P., Müller C., Geyssant J. R., Melières F. and Penn I. E. (1991) Hétérogénéité quantitative et qualitative de la matière organique dans les argiles du Val de Pickering (Yorkshire, U.K.): cadre sédimentologique et stratigraphique. *Rev. Inst. Fr. Pétr.,* **46**, 675-712.

Lallier-Vergès E., Boussafir M., Bertrand P. and Badaut-Trauth D. (1993) Selective preservation of various types of organic matter as assessed by STEM studies on a cyclic productivity-controlled sedimentary series, (Kimmeridge Clay Formation, U.K.). In *Organic Geochemistry.* ØYGARD K. (ed.), Falch Hurtigtrykk, Oslo. pp.384-88..

Largeau C., Derenne S., Casadevall E., Kadouri A. and Sellier N. (1986) Pyrolysis of immature Torbanite and of the resistant biopolymer (PRBA) isolated from extant alga *Botryococcus braunii.* Mechanism of formation and structure of Torbanite. In *Advances in Organic Geochemistry 1985.* LEYTHAEUSER D. and RULLKÖTTER J.(eds.) *Org. Geochem.*, **10**, 1023-1032. Pergamon Press, Oxford.

Largeau C., Derenne S., Casadevall E., Berkaloff C., Corolleur M., Lugardon B., Raynaud J.F. and Connan J. (1990) Occurrence and origin of "ultralaminar" structures in "amorphous" kerogens of various source rocks and oil shales. *Org. Geochem.* **16**, 889-895.

Largeau C. and Derenne S. (1993) Relative efficiency of the Selective Preservation and Degradation Recondensation pathways in kerogen formation. Source and environment influence on their contributions to type I and II kerogens. *Org. Geochem.* **20**, 611-615.

Larter S. R. (1984) Application of analytical pyrolysis techniques to kerogen characterization and fossil fuel exploration/exploitation. In *Analytical Pyrolysis*, VOORHEES K.J.(ed.), Butterworths, London, pp. 212-275.

Le Berre F., Derenne S., Largeau C., Connan J. and Berkaloff C. (1991) Occurrence of non-hydrolysable, macromolecular, wall constituents in bacteria. Geochemical implications. In *Organic Geochemistry. Advances and applications in energy and the natural environment.* MANNING D.A.C. (ed.), University Press, Manchester. pp.428-431..

Leeuw J. W. de and Largeau C. (1993) A review of macromolecular organic compounds that comprise living organisms and their role in kerogen, coal and petroleum formation. In *Organic Geochemistry*. M.H. ENGEL and S.A. MACKO (Eds) Plenum Publishing Corp. pp. 23-72.

Ramanampisoa L.R., Bertrand P., Disnar J. R., Lallier-Vergès E., Pradier B. and Tribovillard N. P. (1992) Etude à haute résolution d'un cycle du carbone organique de roche-mère du Kimméridgien du Yorkshire (G.-B.) : résultats préliminaires de géochimie et de pétrographie organique. *C. R. Acad. Sci., Ser II*, **315**, 1493-1498.

Saiz-Jimenez C. and de Leeuw J.W. (1986) Lignin pyrolysis products: their structures and their significance as biomarkers. In *Advances in Organic Geochemistry 1985* LEYTHAEUSER D. and RULLKÖTTER J.(eds.) *Org. Geochem.*, **10**, 869-876. Pergamon Press, Oxford.

Tegelaar E. W., de Leeuw J. W., Derenne S. and Largeau C. (1989) A reappraisal of kerogen formation. *Geochim. Cosmochim. Acta,* **53**, 3103-3106.

Tissot B.P. and Welte D.H. (1984) Petroleum Formation and Occurrence. Springer, Berlin. 699p.

Van de Meent D., Brown S.C. and Philp R.P. (1980) Pyrolysis-high resolution gas chromatography and pyrolysis gas chromatography-mass spectrometry of kerogens and kerogen precursors. *Geochim. Cosmochim. Acta*, **44**, 999-1013.

Palaeoproduction control on anoxia and organic matter preservation and accumulation in the Kimmeridge Clay Formation of Yorkshire (G.B.): molecular assessment

Jean-Robert Disnar and Lalanirina Ramanampisoa

Université d'Orléans, URA 724 du CNRS, Dépt. des Sciences de la Terre, F-45067 Orléans cedex

Key words- *source-rocks, organic matter accumulation, organic matter preservation, anoxia, Rock-Eval pyrolysis, biomarkers.*

Abstract- In the Cleveland basin, the Kimmeridge Clay Formation (KCF) shows periodical-like fluctuations in the vertical distribution of the organic matter (Herbin *et al.*, 1991). Short-term organic carbon cycles have been studied in detail by Rock-Eval pyrolysis and biomarker analysis. In agreement with other approaches (this volume), the following scheme for organic matter (OM) preservation and accumulation can be proposed from the results of this study. In oxygenated waters, the different biological components of the planktonic OM are subjected to microbial attack in the order of their respective decreasing biodegradability. Globally, the extent of the microbial attack increases with decreasing importance of the originally available stock (i.e. of phytoplanktonic palaeoproduction). The shallow water depth ensures rapid deposition of the planktonic remains with a large part of their original lipid content. During high production periods, high proportions of intracellular pigments and highly metabolisable components are also preserved. In the sediment, the proliferation of anaerobes at the expense of these highly metabolisable compounds results from the development of anoxic conditions which preserved the lipidic hydrocarbon precursors from decay.

Introduction-

Despite the considerable amount of work devoted to oil and/or gas prone source-rocks over recent decades (Tissot and Welte, 1984; Durand, 1988), both their location in sedimentary basins and their conditions of deposition still remain largely obscure (Huc, 1990). A key remaining question is whether organic matter (OM) accumulation is primarily ruled by planktonic production (e.g. Pedersen and Calvert, 1990) or by anoxic conditions of deposition supposed to be especially favorable to OM preservation (e.g. Demaison and Moore, 1980; Bordenave, 1993; Peters and Moldowan, 1993). For the greatest chances of success, the solution to this "anoxia *vs* productivity dilemma" (Pedersen and Calvert, 1990) must be sought first and foremost in series exhibiting rapidly alternating organic matter-poor and -rich layers without pronounced lithological change which could be indicative of marked modifications of the environment of sedimentation. Such features are encountered in the Kimmeridge Clay Formation (KCF) of UK and North Sea (Gallois, 1976; Tyson et al., 1979; Williams, 1988; Oschmann, 1988). Accordingly, the data discussed hereafter have been obtained as part of an extensive study of a 120 cm-long core section of the KCF showing marked and rather regular TOC variations with depth. This core has been sampled on a centimetric scale for detailed investigations on its inorganic and organic components (Ramanampisoa *et al.*, 1992; Tribovillard *et al.*, 1992; Pradier and Bertrand, 1992; other papers presented in this volume). After a short presentation of the main organic geochemistry results already fully discussed in a previous publication (Ramanampisoa and Disnar, 1994), the present paper mainly focuses firstly ideas and concepts on which data interpretation is based and secondly, a synthetic model for OM accumulation in the KCF. Mention will also be made of results obtained on two other core sections of the KCF which have been studied in a similar way (organic geochemistry: Ramanampisoa, 1993; inorganic geochemistry: Tribovillard *et al.*, 1994).

Geological setting, material and methods-

Geological setting-

The lowland plain of the Vale of Pickering strikes at right-angles to the Yorkshire coast of north-eastern England and is floored by Upper Jurassic mudstones, flanked by highlands of older Jurassic limestone to the north and Cretaceous chalk to the south. Sedimentation generally comprises essential clay minerals (kaolinite, illite and mixed-layer illite-smectite) with associated biogenic calcite, mainly planktonic (coccolith) material (Herbin *et al.*, 1991). The Kimmeridgian rocks of Yorkshire have been deposited under low energy depositional environmental conditions (Oschmann, 1988).

Material-

Previous work on the Kimmeridge Clay Formation drilled in the Cleveland basin (Yorkshire, Great-Britain) revealed periodical-like fluctuations of different orders in the vertical distribution of the organic matter (Herbin *et al.*, 1991). The shortest period corresponds approximately to a one metre-thick interval representing ca. 30,000 years. The present study deals with a core section of 120 cm length representative of one single short organic cycle taken from the Eudoxus zone penetrated by the MARTON-87 well. Ninety centimetre-thick samples from this core were first analyzed by Rock-Eval pyrolysis, then 19 of these samples were submitted to classical bitumen extraction and biomarker analysis. A comparable approach has been applied to a second cycle taken higher than the first one from the same well, but also in the *Eudoxus* zone, while a third section had been taken from the *Huddlestoni* zone of the Ebberston 87 well drilled E to the previous one (Ramanampisoa, 1993). These boreholes were drilled in the Vale of Pickering near the Yorkshire coast of north-eastern England (Herbin *et al.*, 1991).

Methods-

Rock-Eval pyrolyses were carried out under standard conditions on a Delsi-Instruments Oil Show Analyzer. Selected samples were extracted with chloroform. The extracts were de-sulfurised on Cu/Zn amalgam and fractionated into saturates, aromatics and polar fractions by liquid chromatography on Florisil. The saturates were analysed by GC and GC-MS. Gas chromatography was carried out on a chromatograph type DI 700 (Delsi-Instruments) equipped with a splitless injector, a FID detector and a 50m x 0.22mm CP Sil 8 CB fused silica capillary column. After splitless injection at 60°C (1 mn), the oven temperature was first increased to 100°C at 30°C/min, then to 290°C at 3°C/min and maintained at the latter temperature until complete elution of all the compounds. Helium was used as carrier gas. GC-MS analyses were conducted on a Varian 300 gas chromatograph equipped with a CP Sil 5 CB (25m x 0.25mm) fused silica capillary column connected to a Finnigan Mat ITD 800 Ion Trap Detector by a 2 m capillary interface heated at 250°C. Molecular organic geochemical parameters were evaluated on the basis of peak areas and relative intensities in mass chromatograms of characteristic ions for regular steranes (m/z 217), 4-methylsteranes (m/z 231), hopanes (m/z 191), n-alkanes and isoprenoids (m/z 85). Tetradec-7-ene was used as an internal standard for GC-MS analysis.

Results and discussion-

Only data necessary to the discussion are presented here. The reader is referred to the paper by Ramanampisoa and Disnar (1994) for detailed organic geochemistry results on the studied carbon cycle.

Organic Matter distribution-

Total organic carbon (TOC) contents decrease rather regularly from a maximum of about 10 % reached near the middle of the core section to about 2 -3 % at its extremities (fig. 1). This peak-shaped TOC variation with depth is characteristic of the short-term carbon cycles corresponding to about 30,000 yrs sedimentation evidenced in the KCF (Herbin *et al.*, 1991). Variations of the Hydrogen Index (HI; mg HC /g TOC) roughly paralleling those of the TOC, reveal that OM geochemical quality improves almost regularly with its contents.

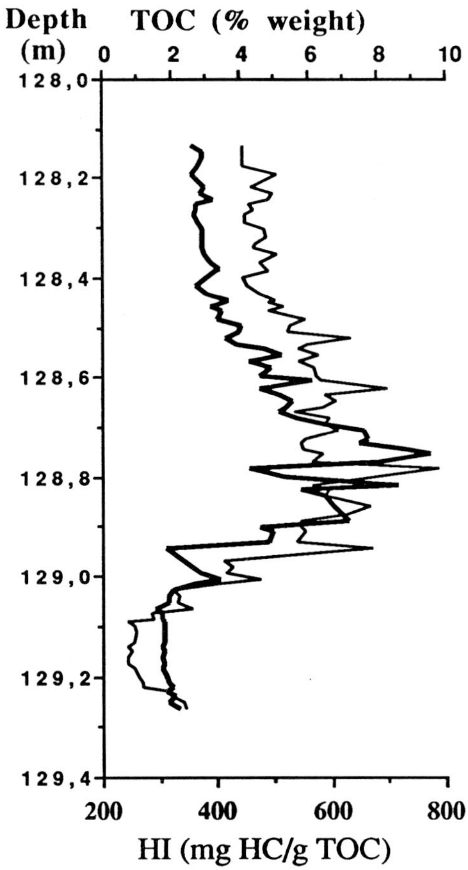

Figure. 1 : Total organic carbon contents (%) and Hydrogen Index (mg hydrocarb. g-1 TOC) variations along the studied KCF section.

Organic Matter maturity-

Tmax values in the 421-431°C range, 60:40 S/R homohopane ratio values and sterane distributions strongly dominated by compounds in C_{27}, C_{28} and C_{29} with the 5a(H),14a(H),17a(H) - 20R stereochemistry indicate a degree of maturity corresponding to the onset of the oil-window.

Origin and preservation of the Organic matter-

(1) Petroleum potential (i.e Rock-Eval S_2 intensity expressed in mg hydroc.g^{-1} rock), (2) bitumen amounts and (3) chlorophyll-derived isoprenoids (pristane + phytane) concentration *vs* TOC regression lines (figs. 2, 3 and 4) all intersect the abscissa axis at

about 1% TOC. This indicates that a large part of the gross qualitative (i.e. HI) variations of the OM originate from the addition of variable amounts of algal-derived hydrogen-rich OM (HI ≈ 700mg HC/g TOC) to a background of about 1 % OM devoid of significant petroleum potential. The latter very likely corresponds to a rather regular supply of terrestrial organic matter clearly evidenced in the palynofacies assemblages (Ramanampisoa *et al.*, 1992) and particularly to inertinite identified on polished rock sections (Williams and Douglas, 1985; Pradier and Bertrand, 1992). Terrestrial plant inputs are also denoted by the presence of low amounts of high-molecular-weight n-alkanes with a well-marked odd over even predominance in the n-C_{20}+ range (Ramanampisoa and Disnar, 1994). However, the high proportions of short-chain n-alkanes with a distribution mode in the n-C_{15} - n-C_{19} range reveals the predominantly algal origin of the OM (Simoneit, 1978; Gelpi *et al.*, 1970). The presence of notable proportions of steranes in C_{27}, C_{28} and C_{29} and of methylsteranes is consistent with autochtonous OM production from an a priori well-diversified (micro-) flora. The absence of notable changes in the relative proportions of the various sterane homologues as well as of the methylsteranes all along the study core section, indicates the apparent constancy in the number and the relative abundance of the main contributing organisms through time (Ramanampisoa and Disnar, 1994).

Pristane and phytane are generally thought to derive from the chlorophyll of photosynthetic organisms (Didyk *et al.*, 1978 and references therein). This classical interpretation is here supported by the rather regular increase in the amounts of these compounds with TOC and even more closely with petroleum potential supposed to be closely related to algal OM accumulation (fig. 4; cf. supra). Pr/n-C_{17} and Ph/n-C_{18} ratio values increase progressively with OM amounts up to a value of about 6% TOC, and then remain constant (fig. 5). The explanation for the evolution of these ratio values is to be found in the biological occurrence and fate of the two kinds of compounds considered i.e. isoprenoids and n-alkanes. Chlorophyll which is supposed to be the main source compound of pristane and phytane (Didyk *et al.*, 1978) originally occurs in chloroplasts in the cells of photosynthetic organisms, and thus can easily be freed in the aqueous medium after cell lysis. Consequently, efficient incorporation of chlorophyll to the sediments necessarily requires deposition of well-preserved and/or even intact algal bodies. This possibility mainly occurs during periods of high primary production, when algal biomass production exceeds the need of grazing organisms (Müller and Suess, 1979; Calvert, 1987; Silver and Gowing, 1991). On the contrary, n-alkanes which mainly derive from fatty acids of rather resistant cell membranes (Tegelaar *et al.*, 1989) are less subject to losses during sedimentation. Accordingly, increasing isoprenoid over n-alkane ratio values should reflect increasing preservation of the original algal OM,

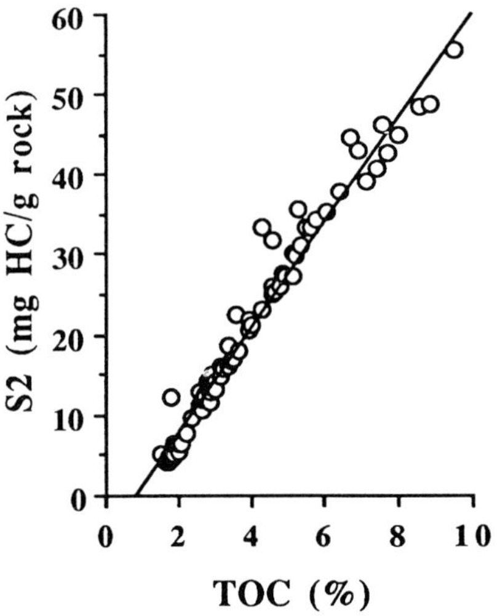

Figure 2 : Petroleum potential (Rock-Eval S2 peak intensity) vs. TOC correlation.

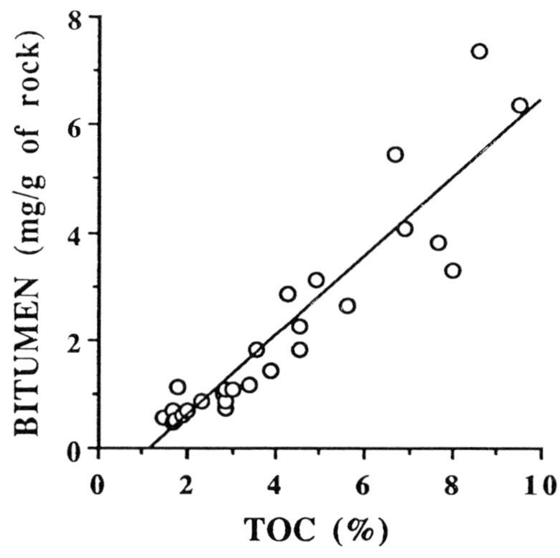

Figure 3 : Chloroform-extracted bitumen amounts vs. TOC (%).

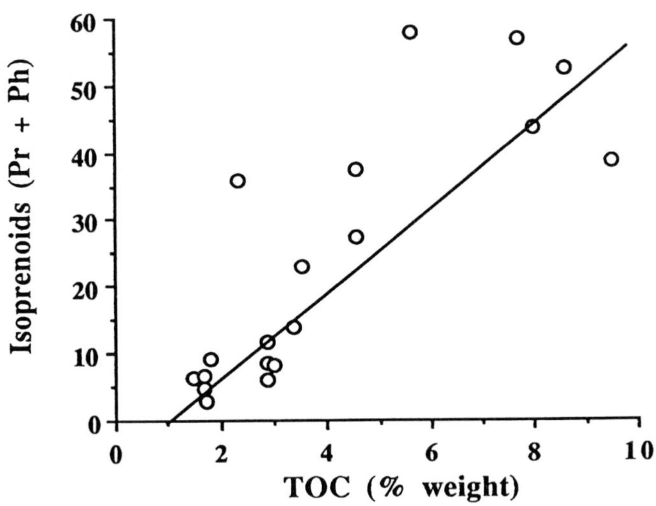

Figure 4 : Variation of the major isoprenoids compounds (Pr + Ph) with TOC contents.

directly related to increasing primary production according to the previous discussion. In addition, when optimum algal OM preservation is reached, i.e. from about 6 % TOC, organic carbon accumulation can be assumed to be directly dependent on primary production.

A comparable deduction can be made from the variations of the methylsterane index (i.e. methylsterane/desmethylsterane ratio): (i) from the premise that precursors of the methylsteranes are much more refractory to biodegradation than those of the desmethylsteranes (Gagosian *et al.*, 1982) and (ii) assuming that there was no notable variation in the relative proportions of the source organisms of these two families of compounds (i.e. a large variety of these for desmethylsteranes and rather specifically dinoflagelates for methylsteranes; de Leeuw and Baas, 1986). Accordingly, (i) decreasing values of this index with increasing TOC observed up to a 6 % value (fig. 5), may be interpreted as a consequence of increasing preservation of the desmethylsterane precursors, while (ii) apparently constant index values observed from about 6 % TOC denote no change in the proportions of the two considered families of compounds and thus their optimum (if not total) preservation. The concordant indications provided by the previous molecular parameters, are, at a gross level, supported and supplemented by HI values which increase regularly with TOC up to about a 6 % value and then tend to stabilise (fig. 5).

Hopane-to-sterane and diasteranes/(diasteranes + regular steranes) ratio values are respectively supposed to denote the variations in the intensity of microbiological activity developed in the sediments and the extent of steroid compounds degradation from the sterene stage of the diagenetic transformation pathway of the original sterol compounds. Both these ratio values show only slight variations along the studied section (factor 1 to 2; fig. 6) compared to those (factor of 5) experienced by the isoprenoids over n-alcane ratios and "Methylsterane index" discussed hereabove (fig. 5).

The absence of significant variations of the Pr/Ph ratio along the considered cycle and its rather high values (≈ 1,9) are in favor of constantly oxic conditions of sedimentation during deposition. Inorganic geochemical data support this point of view (Tribovillard *et al.*, 1992; 1994). On the other hand the decrease of the Pr/Ph ratio values from the first to the second cycle of the Marton 87 core and further from the latter to the third cycle from the Ebberston 87 core could be hypothetically explained by increasing phytane production from methanogens (Ramanampisoa, 1993). This in turn would also indicate increasing anoxia but probably in the sediments since it is welll known that biological methane genesis occurs only after complete sulphate reduction. This

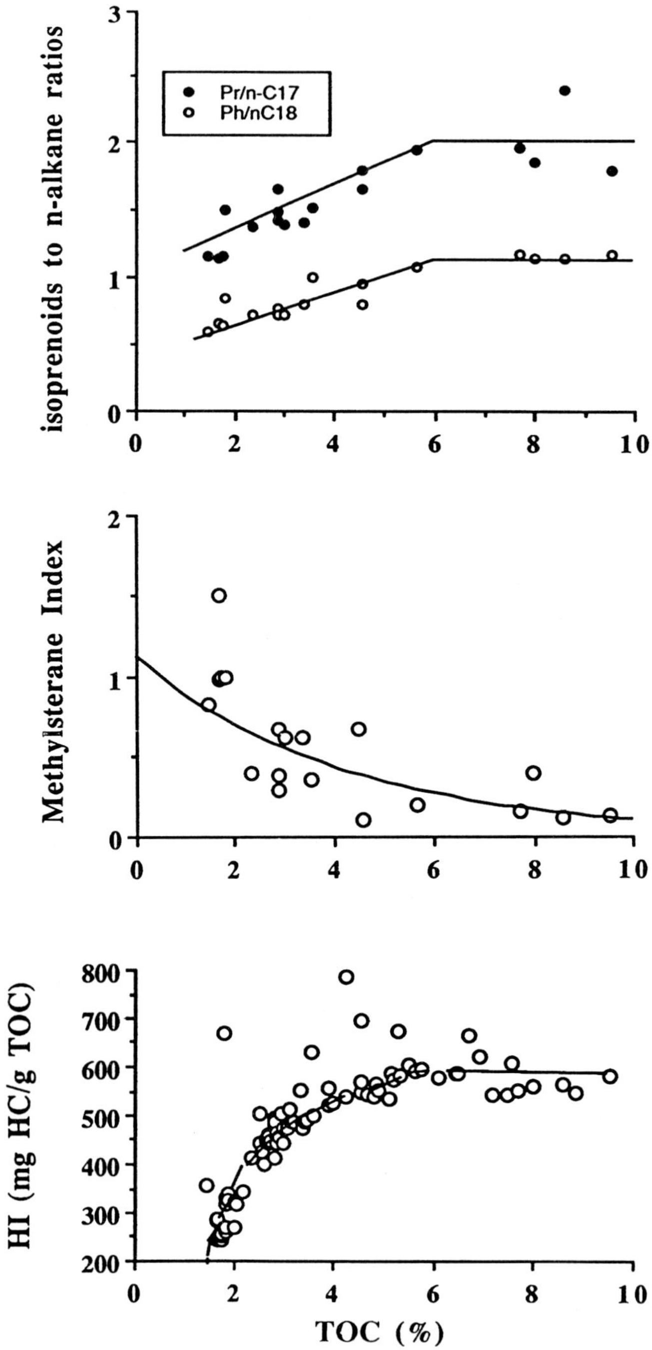

Figure 5 : Variation of (i) isoprenoids-to-n-alkane ratios, (ii) methylsterane index and (iii) HI values with TOC, showing the major change in OM quality intervening from about 6 % TOC.

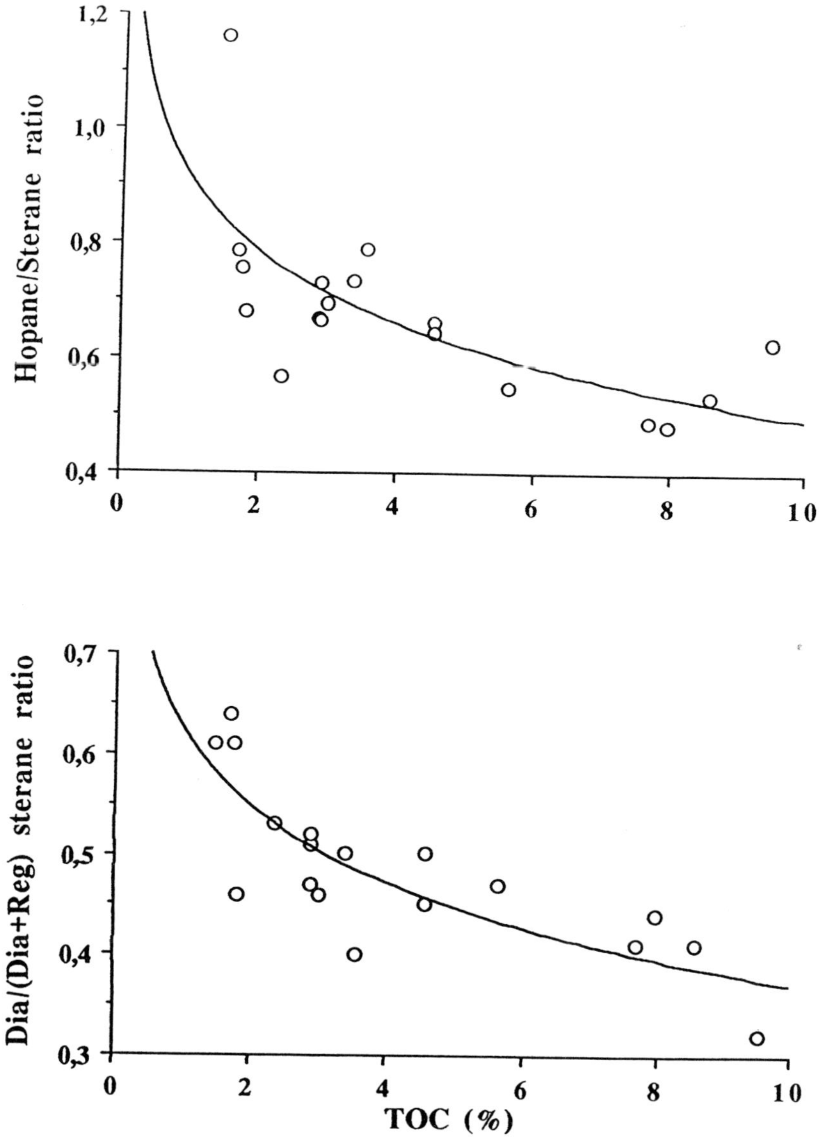

Figure 6 : Variation of (i) the hopane/sterane and (ii) C-27 sterane DIA/(DIA + REG) (rearranged upon rearranged plus regular sterane) ratios with TOC.

explanation is substantiated by TOC contents increasing from cycle 1 to cycle 2 and from the latter to stratigraphic section of the Marton 87 borehole corresponding to cycle 3 in the Ebberston 87 well (Herbin *et al*., 1991).

General discussion-

A main limitation of biomarker analysis is that it is addressed to hydrocarbons derived from original pigment molecules (chorophylls, carotenoids...) or lipidic compounds (fatty acids, sterol hopanoids) well-known for their rather great resistance to biodegradation. No information can be obtained about the possible preservation - if any - of derivatives of highly metabolisable compounds such as proteins and polysaccharides (Cowie and Hedges, 1992), due to the fact that these are composed with too light monomers to produce compounds in the classically examined C_{15+} - C_{35} molecular range. However, differences in the lability and/or biodegradability of hydrocarbon precursors can be used to estimate losses undergone by the original planktonic material during sedimentation and subsequent diagenesis. According to the previous discussion, such differences can originate from intrinsic molecular properties of the various considered compounds (e.g. methylsterols *vs* sterols, rearranged steroids *vs* regular steroids) and/or to the mode of occurrence of the precursor compounds in the biological cell (e.g. chlorophyll in intracellular chloroplasts easily lost after cell lysis *vs* fatty acids in cell membranes highly contributing to kerogen "formation" (Tegelaar *et al.*, 1989)

Previous work evidenced considerable fluctuations in the vertical distribution of organic matter in the KCF (e.g. Herbin *et al.*, 1991). These variations are very probably due to periodical climatic changes (Dunn, 1974; House, 1986). However, neither the exact role played by the climate, nor the major mechanism governing OM accumulation (namely, the part played by primary production and/or anoxia; cf introduction) are yet presently understood (Herbin *et al.*, 1993). Together with other works, the results of the present study on short-term carbon cycles allows the following scheme for OM accumulation in the KCF to be proposed:- During periods of high primary production, a large part of the phytoplankton escaped grazing (Smetacek *et al.*, 1978; Walsh, 1983; Silver and Gowing, 1991). This and a low water depth (≈ 50 m; Oschmann, 1988) allowed intact and/or little altered cells to reach the sediment with intact membrane components (i.e. fatty acids, sterols) and even intracellular material (e.g. chloroplasts, proteins...) easily lost after cell lysis. In the sediment, microbial activity (*a priori* mainly due to microbial sulphate reducers; Bertrand and Lallier-Vergès, 1993) developed at the expense of easily degradable material - probably constituted with proteins and glucides - and left the lipidic components and the pigments almost completely unaltered. Such conditions are reflected by high isoprenoids-over-n-alkane ratio values, low methylsterane indexes and high IH values, all reached from 6 % TOC in the considered KCF carbon cycle (present sediment TOC concentration). Above this TOC level, the constant values of all these parameters (this study) opposed to pyrite contents increasing regularly with TOC (Bertrand and Lallier-Vergès, 1993) can consistently be attributed to enhanced microbial sulphate reduction induced by increasing inputs of easily metabolisable OM to the sediment. Conversely, below 6 % TOC, the regular decrease of the above-mentioned molecular and gross (IH) parameters with decreasing TOC, provides clear evidence for increasing chlorophyll loss and alteration of the original lipids in the water column. The very low amounts of highly metabolisable OM which then reached the sediment only permitted limited sulphate reduction, reflected by low pyrite contents (Bertrand and Lallier-Vergès, 1993). During these low production periods, the inputs of highly metabolisable OM necessary to maintain bacterial sulphate reduction may have been ensured by rapid sedimentation of organisms having a mineral test (e.g. coccoliths; Bertrand and Lallier-Vergès, 1993; also see Cowie and Hedges, 1992). Upward diffusion of H_2S from OM-rich levels could possibly also help to keep anoxia in the uppermost sediment layer.

The absence of significant degradation of lipidic components in the sediments is indicated by the nearly constant value of the diasterane/(dia.+ reg.)-sterane ratio all over the large range of TOC values examined. The seldom observed bisnorhopane (Seifert *et al.*, 1978; Peters and Moldowan, 1993) which is, in the KCF, exclusively but sometimes in high proportions found in TOC rich levels (Ramanampisoa, 1993), very probably originates from a special strain of anaerobic bacteria which could only thrive at the expense of highly metabolisable OM. These observations are consistent with anoxia-induced preservation of the lipidic hydrocarbon precursors in the sediments.

The absence of significant change of the Pr/Ph ratio and its rather high values (≈ 1.9) are in favor of stable aerobic conditions of sedimentation, also supported by inorganic geochemistry data (Tribovillard *et al.*, 1992; 1994). The development of anoxia in the water column is also inconsistent with the low water depth because (1) of supposedly efficient water oxygenation by wave action and (2) the dramatic consequences the poisoning of the photic zone by H_2S would have had on primary production.

Concluding remarks-

Globally the proposed scheme is consistent with the successive microbial attacks, occurring in oxygenated waters, of the biological components of the planktonic OM in the order of their decreasing biodegradability. The intensity of this microbial degradation increases with decreasing importance of the originally available stock (i.e. of planktonic palaeoproduction). The low water depth and the rather large size of the organic debris due to limited degradation of the phytoplankton cells ensured their rapid deposition with a large part of their original lipid content and even, during high production periods, with high proportions of intracellular pigments and highly metabolisable components. In the sediment, the proliferation of anaerobes thriving only at the expense of these highly metabolisable compounds results from the development of anoxic conditions preserving the lipidic hydrocarbon precursors from aerobic decay. Anoxia obviously appears as a determining factor in OM preservation, but develops in the sediment only as a consequence of (i) high primary production and (ii) environmental conditions favorable to limited OM degradation in the water column, i.e. mainly to a low water depth in the considered KCF case. The proposed interpretation is fully consistent with the idea "that source and molecular-level composition are (...) important factors in sedimentary OM preservation" (Cowie and Hedges, 1992).

Acknowledgements- All participants of this research group are gratefully acknowledged for their scientific, technical and/or financial support(s).

References-

Bertrand P. and Lallier-Vergès E. (1993) Past sedimentary organic matter accumulation and degradation controlled by productivity. *Nature,* **364**, 786-788.

Bordenave M.L.(1993) The sedimentation of organic matter. *In* Applied Petroleum Geochemistry. BORDENAVE M.L. (ed.), Editions Technip, Paris. pp. 15-76..

Calvert S.E. (1987) Oceanographic controls on the accumulation of organic matter in marine sediments. *In* Marine Petroleum Source Rocks. BROOKS J. and FLEET A.J. (eds.), Geol. Soc. Spec. Pub., **26,** 137-151.

Cowie G.L. and Hedges J.I. (1992) The role of anoxia in organic matter preservation in coastal sediments: relative stabilities of the major biochemicals under oxic and anoxic depositional conditions. *In*: Advances in Organic Geochemistry. *Org. Geochem.*, **19**, 229-234.

Demaison G.J. and Moore G.T. (1980) Anoxic environments and oil source bed genesis. *AAPG Bull*., **64,** 1179-1209.

Didyk B.M., Simoneit B.R.T., Brassell S.C. and Eglinton G. (1978) Organic geochemical indicators of palaeoenvironmental conditions of sedimentation *Nature,* **272**, 216-222.

Dunn C.E. (1974) Identification of sedimentary cycles through Fourier analysis of geochemical data. *Chem. Geol.,* **13**, 217-232.

Durand B. (1988) Understanding of hydrocarbon migration in sedimentary basins (present state of knowledge). *In* Advances in Organic Geochemistry 1987. MATAVELLI L. and NOVELLI L. (eds.). Pergamon Press, Oxford. *Org. Geochem.*, **13**, 445-459.

Gagosian R.B., Smith S.O. and Nigrelli G.E. (1982) Vertical transport of steroid alcohols and ketones measured in a sediment trap experiment in the Equatorial Atlantic Ocean. *Geochim. Cosmochim. Acta,* **46**, 1163-1172.

Gallois R.W. (1976) Coccolith blooms in the Kimmeridge Clay and the origin of North Sea Oil. *Nature*, **259**, 473-475.

Gelpi E., Schneider H., Mann J. and Oro T. (1970) Hydrocarbons of geochemical significance in microscopic algae. *Phytochem*., **9**, 603-612.

Herbin J.P., Müller C., Geyssant J.R., Mélières F., Penn I.E. and the group Yorkim-IFP (1991) Hétérogénéité quantitative et qualitative de la matière organique dans les argiles du Kimmeridgien du Val de Pickering (Yorkshire, UK): cadre sédimentologique et stratigraphique. *Rev. Inst. Fr. Pétrol.,* **46**, 675-712.

Herbin J.P., Geyssant J.R., El Albani A., Colbeaux J.P., Deconink J.F., Fernandez-Martinez J.L., Proust J.N. and Vidier J.P. (1993) Sequence stratigraphy of source rocks applied to the study of the Kimmeridgian/Tithonian in the Northwestern European Shelf (Dorset/UK, Yorshire/UK and Boulonnais,/France). *Mar. & Pet Geol.*, in press.

House M.R. (1986) Are Jurassic sedimentary microrhythms due to orbital forcing? *Proceedings of the Ussher Society,* **6**, pp. 299-311.

Huc A.Y. (1990) Understanding organic facies: a key to improved quantitative petroleum evaluation of sedimentary basins. *In*: Deposition of organic facies. HUC A.Y. (ed.), AAPG studies in geology 30, AAPG, Tulsa, pp. 1-11.

Leeuw J.W. de and Baas M. (1986) Early diagenesis of steroids. *In* Biological markers in the sedimentary record. JOHNS R.B .(ed.), Elsevier, Amsterdam. pp. 101-123.

Müller P. and Suess E. (1979) Productivity, sedimentation rate and sedimentary organic matter in the oceans I. Organic carbon preservation. *Deep-Sea Res.,* **26**, 1347-1362.

Oschmann W. (1988) Kimmeridge Clay Sedimentation - A new cyclic model. *Palaeogeogr., Palaeoclimatol., Palaeoecol.,* **65**, 217-251.

Pedersen T.F. and Calvert S.E. (1990) Anoxia *vs*. productivity: What controls the formation of organic-carbon-rich sediments and sedimentary rocks? *AAPG Bull.*, **74**, 454-466.

Peters K.E. and Moldowan J.M. (1993) The biomarker guide. Prentice Hall, Englewood Cliffs. 363 p.

Pradier B. and Bertrand P. (1992) Etude à haute résolution d'un cycle du carbone organique de roche-mère du Kimméridgien du Yorkshire (G.B.): relation entre composition pétrographique du contenu organique observé *in situ*, teneur en carbone organique et qualité pétroligène. *C. R. Acad. Sci. Paris*, **315**(2), 187-192.

Ramanampisoa L. (1993) Etude des mécanismes responsables de l'accumulation de la matière organique dans les argiles du Kimméridgien- Yorshire (UK): approches géochimiques et pétrographiques organiques. Thèse, Université d'Orléans, 220p.

Ramanampisoa L.and Disnar J.R. (1994) Primary control of paleoproduction on organic matter preservation and accumulation in the Kimmeridge rocks of Yorshire (UK). *Org. Geochem.* , **21**-12, 1153-1167.

Ramanampisoa L., Bertrand P., Disnar J.R., Lallier-Vergès E., Pradier B.and Tribovillard N.P. (1992) Etude à haute résolution d'un cycle de carbone organique des argiles du Kimméridgien du Yorkshire (G.B.): résultats préliminaires de géochimie et de pétrographie organique. *C R. Acad. Sci. Paris*, **315**(2), 1493-1498.

Seifert W.K., Moldowan J.M., Smith G.W. and Whitehead E.V. (1978) First proof of a C_{28}-pentacyclic terpane in petroleum. *Nature,* **271**, 436-437.

Silver M.W. and Gowing M.M. (1991) The "particle" flux: origins of biological components. *Prog. Oceanog.*, **26**, 75-113.

Simoneit B.R.T. (1978) Organic geochemistry of terrigeneous muds and various shales from the Black Sea, DSDP Leg 42b, *In*: Initial Reports of the Deep Sea Drilling Project-XLII-2. ROSS D.A., NEPROCHNOV Y.P. *et al*., (eds.), **42**-2, 749-753.

Smetacek V., Brockel K.V., Zeitschel B. and Zenk W. (1978) Sedimentation of particulate matter during a phytoplankton spring bloom in relation to hydrographical regime. *Mar. Biol.,* **47**, 211-226.

Tegelaar, E.W., de Leeuw J.W., Derenne S. and Largeau C. (1989) A reappraisal of kerogen formation . *Geochim. Cosmochim. Acta* , **53**, 3103-3106.

Tissot B.P. and Welte D.H (1984) Petroleum Formation and Occurrence, 2nd Ed., Springer Verlag Berlin, 699 pp.

Tribovillard N.P., Desprairies A., Bertrand P., Lallier-Vergès E., Disnar J.R. and Pradier B. (1992) Etude à haute résolution d'un cycle du carbone organique de roche-mère du Kimméridgien du Yorkshire (G.B.): minéralogie et géochimie (résultats préliminaires). *C. R. Acad. Sci. Paris*, **314**(2), 923-930.

Tribovillard N.P., Desprairies A., Bertrand P., Lallier-Vergès E., Bertrand P., Moureau N., Ramdani A. and Ramanampisoa L. (1994) Geochemical study of organic matter-rich cycles from the Kimmeridge Clay formation of Yorshire (UK): productivity versus anoxia. *Palaeogeogr. Palaeoclim. Palaeoecol*., **108**, 165-185.

Tyson R.V., Wilson R.C. and Downie C. (1979) A stratified water column environmental model for the type Kimmeridge Clay. *Nature*, **277**, 377-380.

Walsh J.J. (1983) Death in the sea: enigmatic phytoplankton losses. *Prog. Oceanog.*, **12**, 1-86.

Williams P.F.V. (1986) Petroleum Geochemistry of the Kimmeridge Clay of Southern and Eastern England . *Mar. & Pet. Geol*., **3**, 258-281.

Williams P.F.V. and Douglas A.G. (1985) Organic geochemistry of the British Kimmeridge clay. 1: composition of shale oil produced from Kimmeridge sediments. *Fuel*, **64**, 1062-1069.

Clay diagenesis in organic-rich cycles from the Kimmeridge Clay Formation of Yorkshire (G.B.): implication for palaeoclimatic interpretations

Alain Desprairies, Mostafa Bachaoui, Abdelkader Ramdani and Nicolas Tribovillard

Université Paris Sud, URA 723 du CNRS, bâtiment 504, F-91405 Orsay cedex

Key-words- *clay diagenesis, organic-rich sediments, Kimmeridge Clay Formation, geochemical cyclicity, palaeoclimatic indicators.*

Abstract- Mineralogical and geochemical analyses have been performed on organic matter-rich mudstone sections from the Kimmeridge Clay Formation cored in the Cleveland basin (Marton and Ebberston boreholes) onshore Yorkshire. The formation shows a hierarchical cyclicity in organic matter distribution. Three short-term cycles (1 metre in thickness) having various organic carbon contents, have been studied with a high resolution sampling to estimate qualitatively and quantitatively the diagenetic transformations occurring in the clay fraction of sediments. Electron microscopy results and mass-balance calculations indicate a homogeneous degree of diagenetic reactions (kaolinitisation of micas, illitisation of smectite) never definitively oblitering the nature of primary flux of detrital clay minerals. Consequently, the distribution of clay mineral assemblages (as kaolinite *versus* mixed-layer illite-smectire or smectite) combined with correlative variations of major and trace element parameters (such as Th/K ratio) has been used throughout each sedimentary cycle as indicators of compositional changes of terrestrial inputs. Climatic oscillations from more aridic conditions to warmer and more humid ones appear to be related to the third-order cyclicity. A marked coincidence between the high accumulations of organic carbon, controlled by primary production, and the increase of land weathering which controls the nutriment fluxes, has also been emphasised.

Introduction-

The Jurassic Kimmeridge Clay Formation (KCF) from southern England displays a lithology made of alternating organic-rich and calcareous mudstones (Cox and Gallois, 1981). Changes in detrital clay, carbonate and organic inputs appear to depend on environmental factors; facies variations visible along the depositional sequence of the KCF present a cyclic character. Assuming sedimentation rates between 0.1 mm per year (shales) and 1mm per year (coccolith limestones), Oschmann (1988) distinguished four different types of cycles, whose duration ranges from one year (seasonal four-order cycles) to 3.10^7 years (first-order cycles). Third-order cycles (5.10^3 to 15.10^3 years long) are the easiest to recognise on field, with facies changing on a decimetric to metric scale, from marly shales, shales, bituminous shales and coccolith limestones and *viceversa*. According to Oschmann (1988), major and minor sea-level fluctuations provide the best interpretation for first- and second-order cycles, whereas third- and fourth-order cycles appear to be mostly controlled by climatic variations.

The vertical distribution of organic matter (OM) over the KCF and its heterogeneity, both qualitative and quantitative, also enabled to identify hierarchical cyclic events. Continuously cored boreholes in the Cleveland Basin (Yorkshire) have shown at least two orders of cyclicity, with megacycles grouping elementary cycles expressed through the total organic carbon (TOC) distribution (Herbin *et al.*, 1991; 1993). The *mean* duration of elementary cycles, approximately 25.10^3 years, and their organised lithological variations coupled with their TOC content, are common characteristics of the third-order cycles defined by Oschmann (1988). Detailed petrographical and geochemical analyses (organic and inorganic) have been performed on the elementary cycles in order to decipher which favourable factors, bottom water oxygen concentration or surface water productivity besides sedimentation rates, were to be taken into account to explain OM entrapment and preservation (Pradier and Bertrand, 1992; Ramanampisoa *et al.*, 1992; Bertrand and Lallier-Vergès, 1993; Tribovillard *et al.*, 1992, 1994).

To date, however, no clay mineralogy study has been done at a high resolution scale through elementary cycles (i.e. third-order ones). The aim of this paper is to assess the possible modification affecting clay assemblages during diagenesis in order to test their significance as environmental, climatic, indicators.

Sampling and methods-

Samples were recovered from two boreholes (Marton 87 and Ebberston 87) cored by the "British Geological Survey" and the "Institut Français du Pétrole" in the Cleveland Basin of Yorkshire (fig. 1). From the Marton 87 borehole, two elementary cycles were chosen near the top of the Eudoxus ammonite zone (fig. 2). The cyclicity is expressed through variations in the nature and abundance of organic matter (Herbin *et al.*, 1991; Ramanampisoa *et al.*, 1992). The first cycle chosen, called Marton I, shows a TOC content varying between 1 and 8%. TOC values within the second cycle chosen, called Marton II, fluctuate between 5 and 29%. In the Ebberston borehole, the third cycle studied, Ebberston I, is located at the top of the *Wheatleyensis* ammonite zone. TOC values vary between 3 and 18%.

Marton I (120 cm long) was split into 90 samples; Marton II (62 cm long), was split into 47 samples and Ebberston I (2 metres long) into 25 samples.

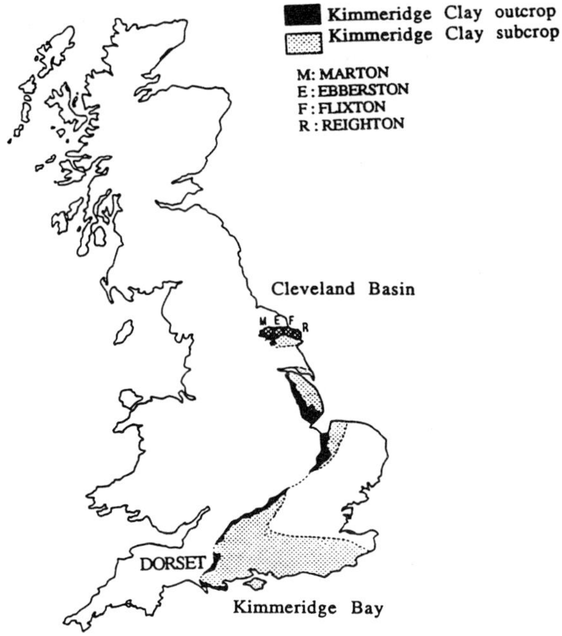

Fig. 1 - Location of the four boreholes cored in the Cleveland Basin (onshore Yorkshire) : Marton, Ebberston, Flixton, Reighton (after Herbin *et al.*, 1991).

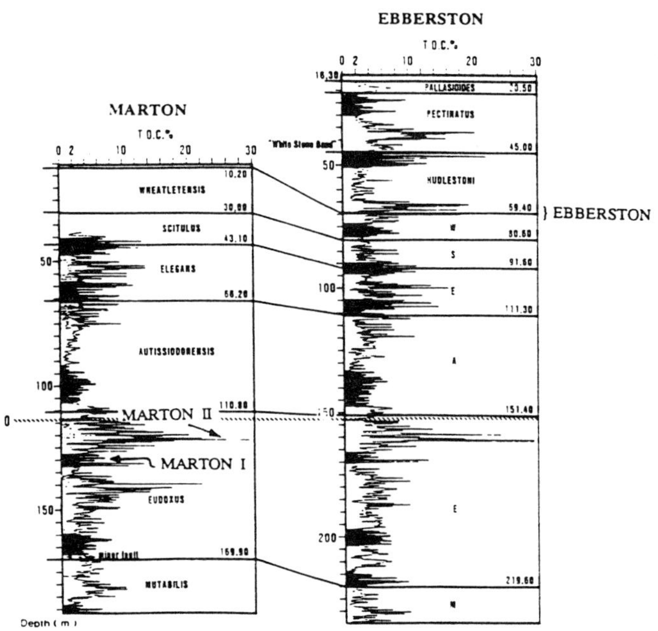

Fig. 2 - Total organic carbon (TOC) distribution with depth in Marton and Ebberston boreholes. Location of the elementary, (after Herbin *et al.*, 1991) i.e. third-order, cycles studied.

From X-ray diffraction (XRD) diagrams, quantitative estimations of clays and non-clay minerals (quartz, feldspar, carbonate, pyrite) were carried out on every sample crushed into powder. Clay mineral identification was made on the < 2 µm fraction of 32 samples, previously selected for geochemical studies (Tribovillard *et al.*, 1992, 1994). Percentages of the main clay minerals identified (kaolinite, mica, mixed-layer illite/smectite) were determined from peak heights or area of diffractograms from oriented sample mounts, after glycolation. Infrared spectrometry analyses were also performed in order to control the validity of this semi-quantitative measurement and, more particularly, to improve the estimation of kaolinite abundance.

Bulk-rock chemical analyses were made on each sample, using a scanning electron microscope (SEM) equipped with an energy dispersive spectrometer (EDS) system. On 17 selected samples, the morphology and the chemical composition of clay particles were studied with a transmission electron microscope (TEM-EDS).

Results-

Bulk mineralogy-

XRD data from bulk-rock samples are expressed as mineral percentages (fig. 3 to 5). Total clay content (Table1) ranges on average from 55% (Marton II) to 73% (Marton I).

Table 1 - X-ray diffraction data from the three short-term cycles studied. Abundance of mineralogical components is expressed in weight %.

	Clay	Quartz	Calcite	Gypsum	Pyrite	Feldspars
MARTON I (Nb: 86)						
Min %	60.51	7.34	3.63	0,0	0.52	0,0
Max %	81.86	26.35	19.93	1.42	6.21	2.99
Mean	73.17	13.44	10.95	0.15	1.63	0.65
Std deviation	5.52	3.37	3.72	0.31	1.1	0.63
MARTON II (Nb: 42)						
Min %	39.43	7.85	7.19	0,0	1.1	0,0
Max %	72.6	27.79	35.75	14.87	16.12	7.59
Mean	55.22	18.59	17.43	1.06	5.74	1.76
Std deviation	8.56	4.93	7.25	2.26	3.65	1.46
EBBERSTON I (Nb: 50)						
Min %	43.7	4.82	6.73	0,0	0.91	0,0
Max %	80.23	14.54	29.19	5.39	6.47	5.73
Mean	67.2	9.57	14.93	1.11	2.41	1.34
Std deviation	9.83	1.81	0.71	1.07	1.32	1.07

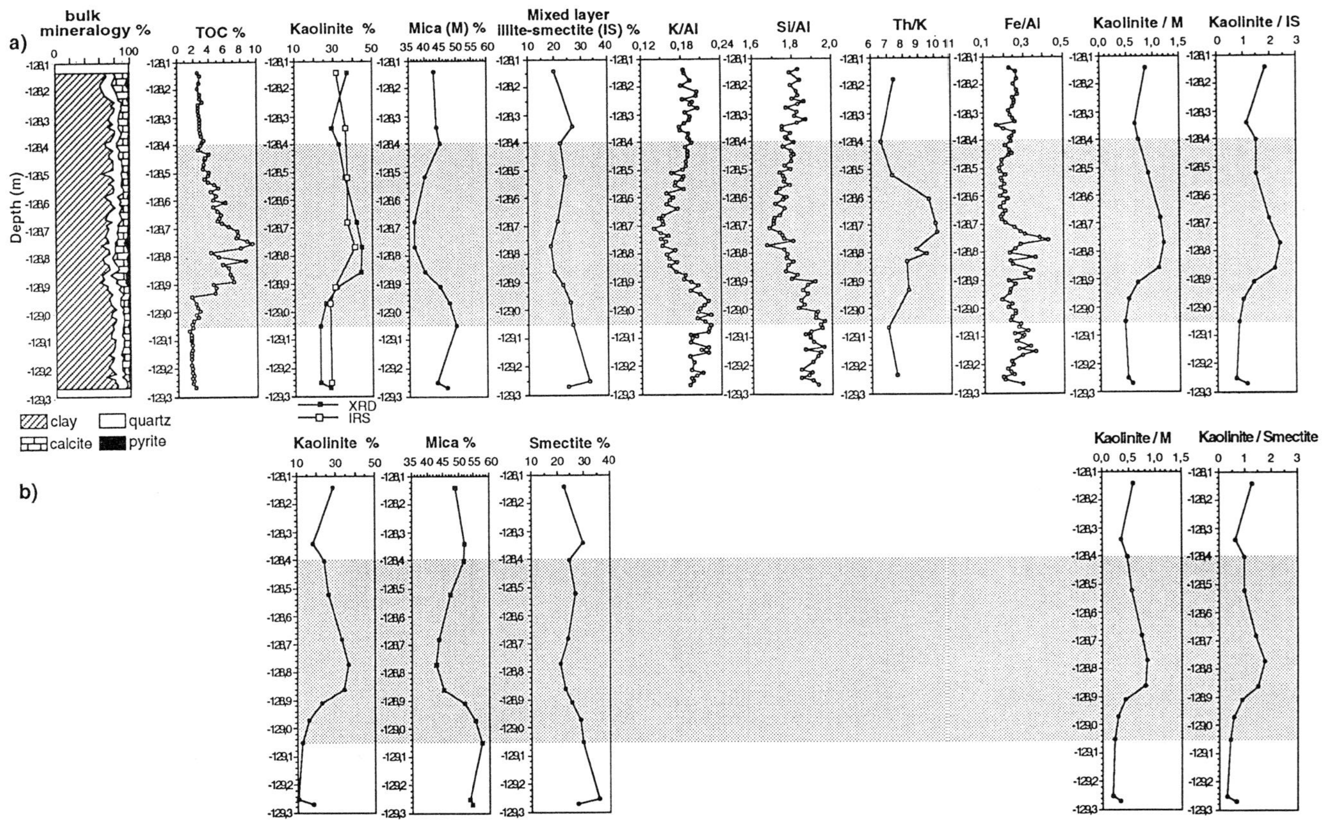

Fig. 3 - **MARTON I cycle**. a - Bulk composition variability and distribution patterns of clay species present throughout a third-order cycle. Comparison between mineralogical (clay mineral ratios) and chemical (elemental ratios) curves with TOC content. b - Profiles of primary terrestrial influx of detrital clay minerals, inferred from the diagenetic model (see text). XRD: X-ray diffraction data. IRS: infra-red spectrometry data.

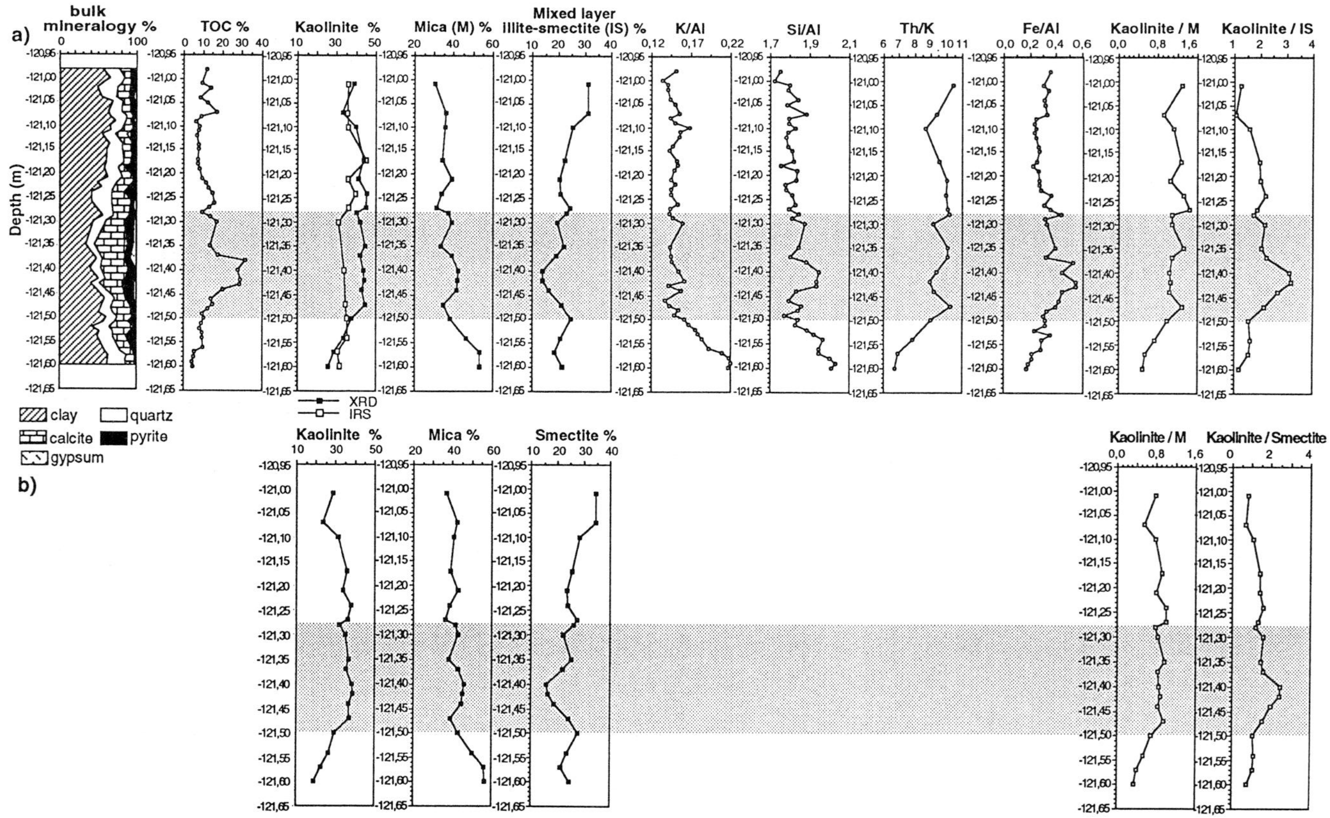

Fig. 4 - **MARTON II cycle**. a - Bulk composition variability and distribution patterns of clay species present throughout a third-order cycle. Comparison between mineralogical (clay mineral ratios) and chemical (elemental ratios) curves with TOC content. b - Profiles of primary terrestrial influx of detrital clay minerals, inferred from the diagenetic model (see text). XRD: X-ray diffraction data. IRS: infra-red spectrometry data.

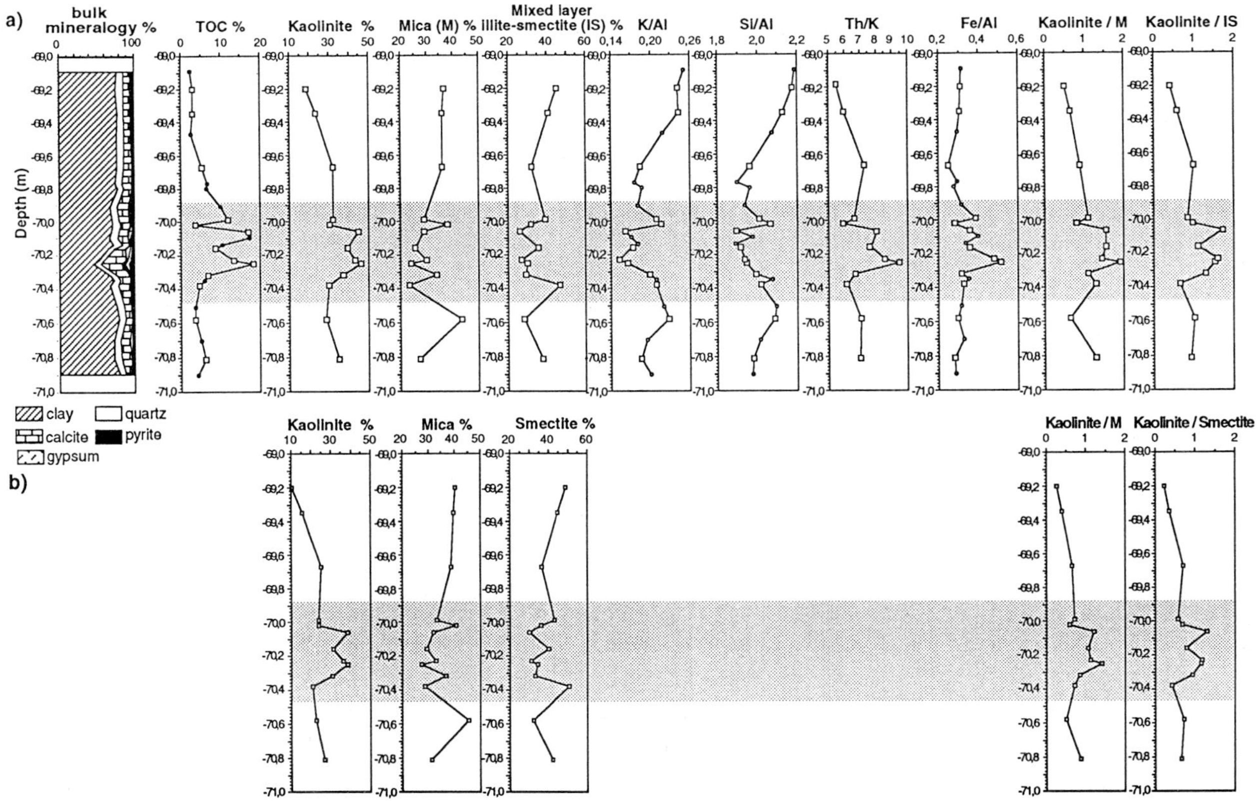

Fig. 5 - **EBBERSTON I cycle**. a - Bulk composition variability and distribution patterns of clay species present throughout a third-order cycle. Comparison between mineralogical (clay mineral ratios) and chemical (elemental ratios) curves with TOC content. b - Profiles of primary terrestrial influx of detrital clay minerals, inferred from the diagenetic model (see text). XRD: X-ray diffraction data. IRS: infra-red spectrometry data.

Quartz is ubiquitous, being present whatever the clay content (10 to 20%). On the other hand, calcite amounts (11 to 18% on average) appear as a dilution factor for the clay-size fraction of the sediment. Feldspars are present sporadically and in low abundance (≤ 2%) and pyrite, following sulphate reduction processes, is strongly correlated with TOC values. Finally, as previously noted (Tribovillard *et al.*, 1992, 1994), the OM content of mudstones does not depend on the lithology.

Clay mineralogy-

The clay mineral assemblage is qualitatively uniform, comprising kaolinite, mica, mixed-layered illite/smectite (IS) as the main phases and chlorite as a minor phase. In the vast majority of samples, using criteria of Reynolds and Hower (1970), IS was interpreted as an ordered interstratified mineral, with a single, somewhat broad, basal peak splitting into 13-14 Å after glycolation. The d-spacing of the 002 reflection indicates a smectite-layer percentage averaging 35%. The second type, illustrated by a small number of samples, displays a basal peak expanding to 17 Å after glycolation, this was identified as randomly interstratified IS, with a smectite layer relative abundance exceeding 50%. Both types may coexist within the same sample.

Clay morphology and chemistry -

With TEM observations under x 5000 to 100,000 magnification, three common morphological textures for platy particles were distinguished:

- anhedral particles of large size (≥ 2 μm) with irregular borders;
- sub-euhedral particles of small size (≤ 0.5 μm) showing equal-dimensional to short-lath-like habitus or pseudohexagonal shape;
- flaky aggregates (0.5 to 1.5 μm in size) of very thin, irregularly shaped, lamellae.

EDS analyses of individual platelets or aggregates indicated three major compositions for clay minerals.

- SiO_2/Al_2O_3 molar ratio close to 2 (ranging from 1.58 to 2.60 and averaging 2.21) and the lack of K_2O evokes the composition of kaolinite (Table 2). These characteristics are found both on anhedral particles of large size (Plate 1) and on well-shaped pseudo-hexagonal flakes (Plate 3). The ferric-oxide content ranges from 0 to 6% and averages 2% regardless of particle morphology or size. Presumably, part of the iron is not present within kaolinite structure and occurs as coating impurities.

Chemical structural formulae have been calculated from EDS analyses (fig. 6) and plotted on the ternary diagram from Velde (1977, 1985), using MR^3, $2R^3$, $3R^2$ coordinates (with MR^3 = Na+K+Ca, $2R^3 = (Al^{3+}+Fe^{3+}+MR3)/2$, $3R^2=(Mg^{2+}+Fe^{2+}+Mn^{2+})/3$). As expected, on this chemiographic representation, kaolinite is found near the $2R^3$ pole (fig. 7).

- SiO_2/Al_2O_3 molar ratio close to 3 (between 2.52 and 3.30, averaging 2.93), high K_2O content (6-7% on average), low (0-5%) or negligible (< 1%) content for Fe_2O_3 and MgO are evidence of mica minerals. Anhedral shape and large size are always the common features of these minerals (Plate 1). The area of mica minerals (fig. 8) plotted in a Velde diagram is located approximately midway on the line joining MR^3 and $2R^3$ (total layer charge between 0.4 and 0.8; fig. 7).

- Some of these mica minerals appear to be altered, bearing discrete lamellae showing sub-euhedral shape (Plate 1) at the surface and at the edge of clear crystals. The close association of this additional clay phase with detrital mica raised problems in EDS analysis. Very occasionally, arrangement into packages of the overgrowing clay crystals permitted to identify the latter as authigenic kaolinite. In most cases, EDS analyses reveal chemical compositions intermediate between those of kaolinite and those of mica: SiO_2/Al_2O_3 molar ratio averaging 2.8, K_2O content ranging from 1.6 to 5.7% (Table 2).

Structural formulae, calculated on an eleven oxygen atom basis, indicated an excess (> 2) of trivalent ions (Al^{3+}, Fe^{3+}) theoretically present in the octahedral sites (fig. 9), allowing the introduction of Al and/or Fe in interlayer position, which is unlikely. On the other hand, no mixed-layer 7-10 Å structure was identified from XRD diagrams. Finally, we think that these analyses, distributed in the Velde diagram between the kaolinite and mica poles (fig. 7) have to be interpreted as a mixture of the latter two minerals, representing *in situ* degradation of detrital micas and correlative kaolinite formation.

- In relation to the chemical composition of micas and to specimens of smectite from the Boulonnais area (see below), SiO_2/Al_2O_3 molar ratio close to 3.5 on average (ranging from 2.8 to 4.5), variable K_2O content (2.4 to 4.7%) are diagnostic values of mixed-layer IS (Table 2, fig. 10). Supporting these data are TEM observations showing platelets growing at the edges of flaky aggregates of thin particles: EDS analyses reveal K_2O content, *i.e.* illite layers, increasing across aggregates from the central area to sub-euhedral platelets (Plate 2). Moreover, according to XRD diffractograms, IS constitute nearly the whole of the ≤ 0.3 µm clay fraction of sediments.

Mica (a)		SiO2	Al2O3	Fe2O3	MgO	CaO	K2O
Nb: 20	Mean %	57.17	32.72	2.29	0.14	0.09	7.59
Marton I & II	Std deviation	3.36	2.05	2.28	0.59	0.29	1.9

Mica (b)		SiO2	Al2O3	Fe2O3	MgO	CaO	K2O
Nb: 20	Mean %	56.29	36.47	2.7	0	0.39	4.13
Marton I	Std deviation	2.73	2.73	1.36	0	0.47	2.17
Nb: 43	Mean %	57.59	33.94	4.82	0.08	0.15	3.42
Marton II	Std deviation	3.4	2.55	2.38	0.23	0.53	2.02

Kaolinite		SiO2	Al2O3	Fe2O3	MgO	CaO	K2O
Nb: 33	Mean %	55.61	42.2	2.18			
Marton I&II	Std deviation	2.78	3.34	2.43			

Mixed layer illite-smectite		SiO2	Al2O3	FeO	MgO	CaO	K2O
Nb: 17	Mean %	59.29	28.19	6.26	1.15	0.98	3.86
Marton I	Std deviation	3.42	3.41	4.26	1.57	0.91	1.1
Nb: 35	Mean %	59.5	29.09	5.27	1.96	0.47	3.69
Marton II	Std deviation	3.07	3.47	2.41	1.53	1.04	1.29
Nb: 14	Mean %	59.67	28.91	6.31	1.2	1.07	2.84
Ebberston I	Std deviation	2.82	2.83	2.37	1.11	0.51	0.89
Nb: 14	Mean %	58.02	30.27	5.66	0.68	0.71	4.65
Dorset	Std deviation	2.85	1.44	2.16	0.76	1.04	1.67
Nb: 20	Mean %	59.49	24.82	6.94	3.26	1.78	4.69
Boulonnais	Std deviation	3.37	3.66	2.37	1.96	0.86	1.22

Smectite		SiO2	Al2O3	FeO	MgO	CaO	K2O
Nb: 4	Mean %	62.5	24.86	5.31	4.25	0.46	2.62
Boulonnais	Std deviation	2.48	3.14	1.53	1.67	0.6	0.77

Table 2 - TEM/EDS chemical analyses performed on detrital and diagenetic clay minerals from Marton and Ebberston boreholes and from outcrops in the Dorset and Boulonnais areas. Mica : a - clear crystals, b - particles degraded "in situ" into kaolinite.

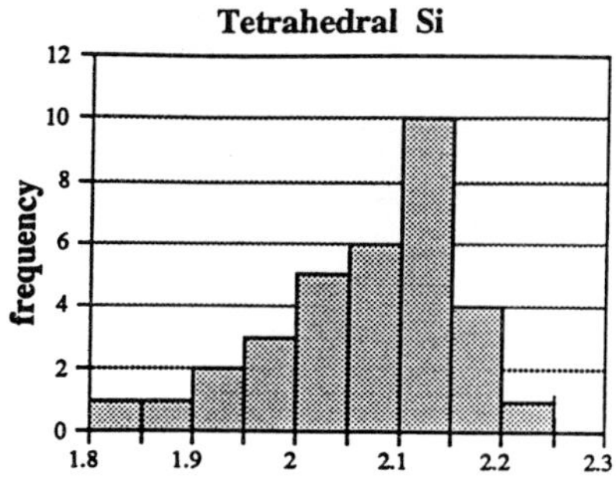

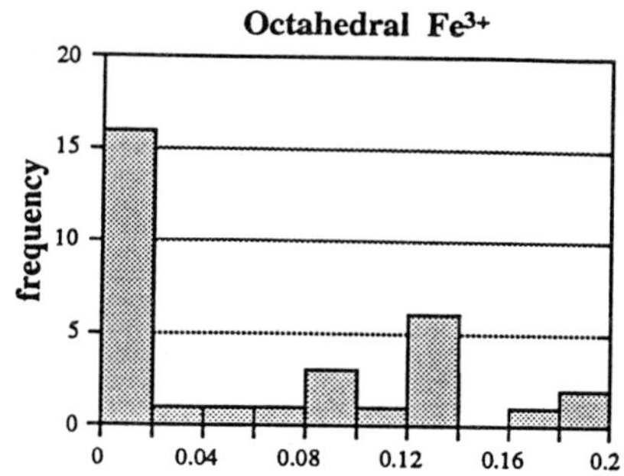

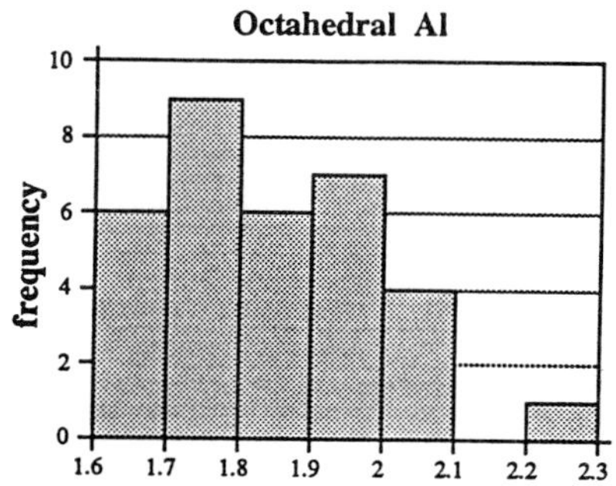

Fig. 6 - Histograms showing the distribution of cations in the structural formulae of 33 kaolinite particles from Marton I & II.

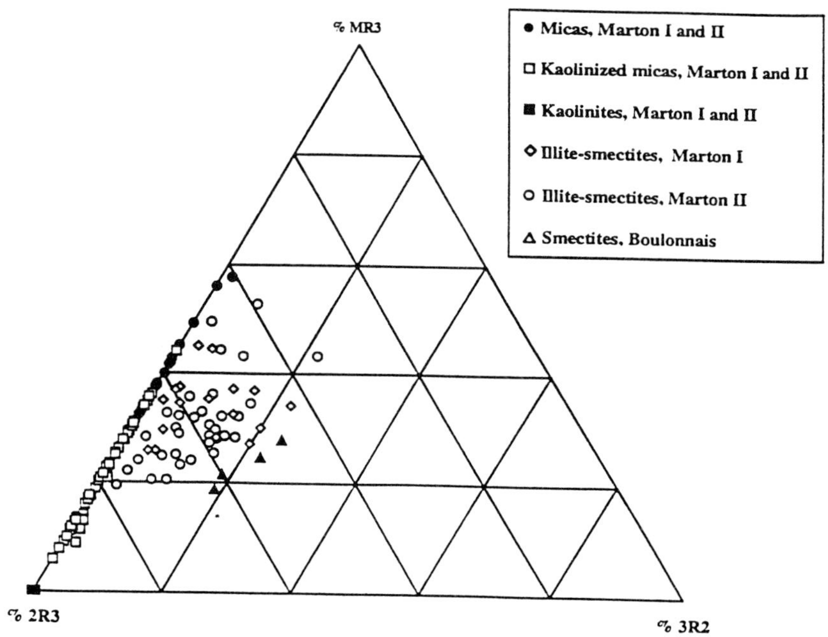

Fig. 7 - Distribution of the cations of clay minerals plotted on a Velde diagram where $MR^3 = K + Na + 2Ca$, $2R^3 = (Al + Fe^{3+} - MR^3)/2$ and $3R^2 = (Mg + Fe^{2+} - Mn)/3$.

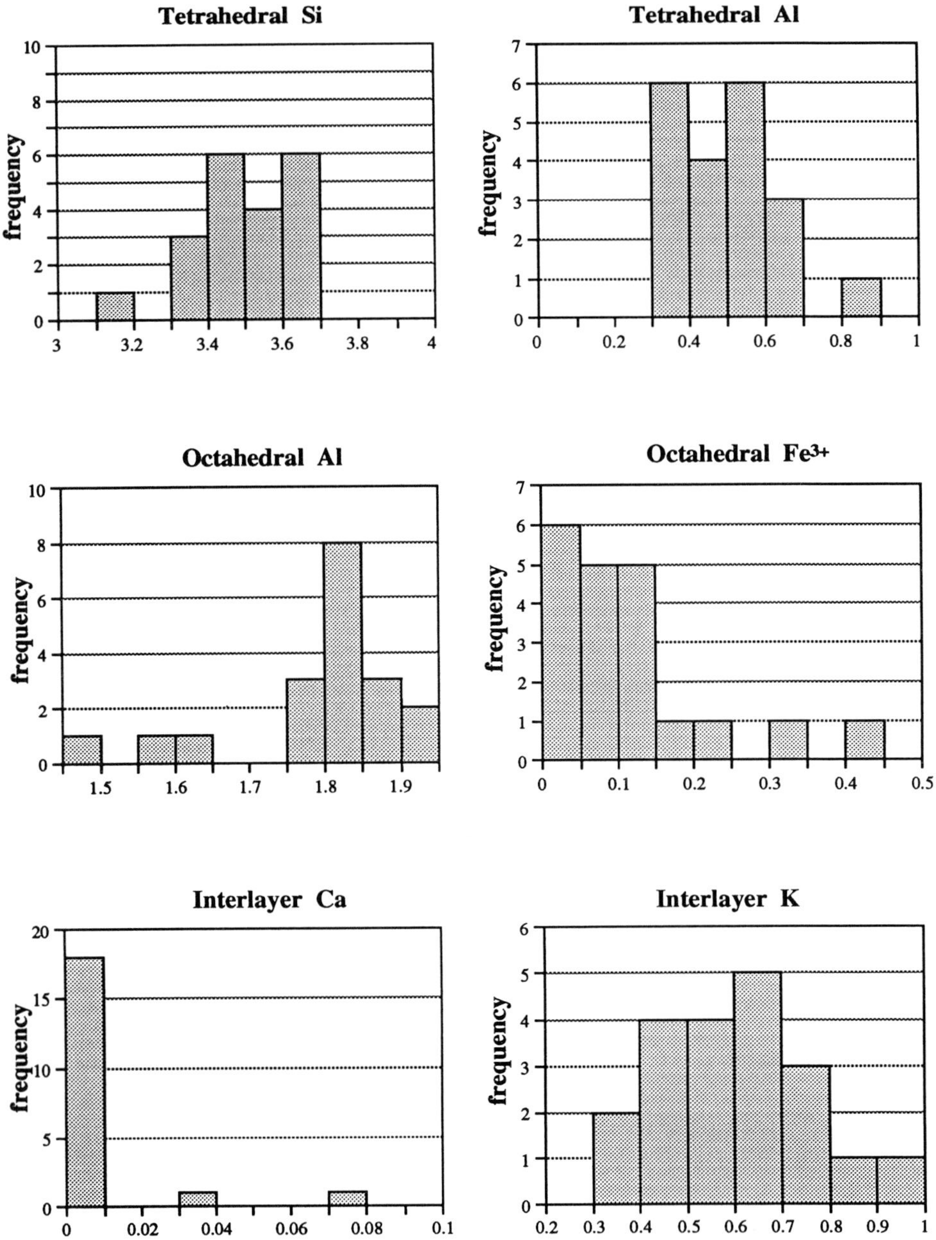

Fig 8 - Histograms showing the distribution of cations in the structural formulae of 20 mica particles from Marton I & II.

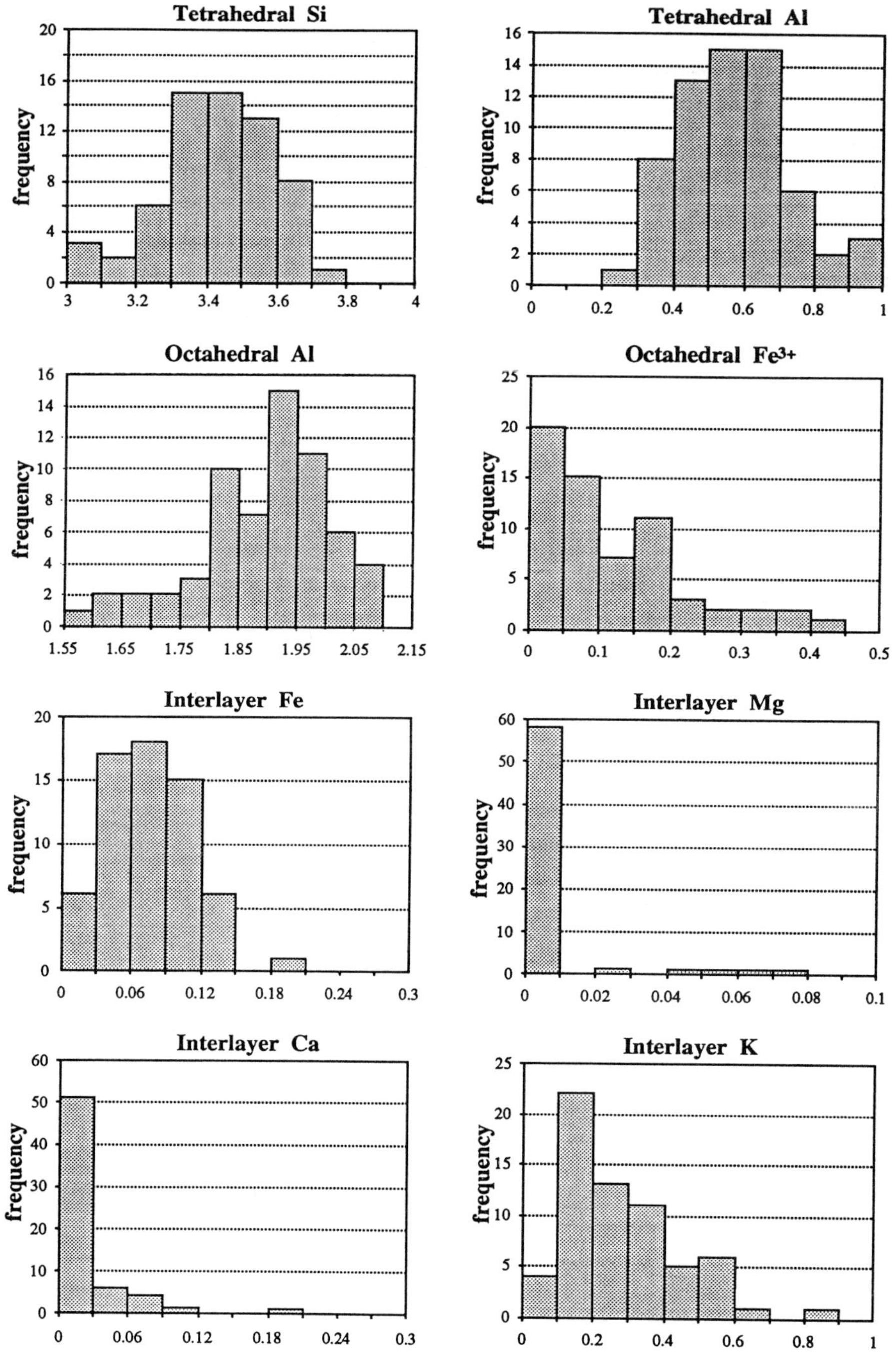

Fig. 9 - Histograms showing the distribution of cations in the structural formulae of 63 kaolinitized mica particles from Marton I & II.

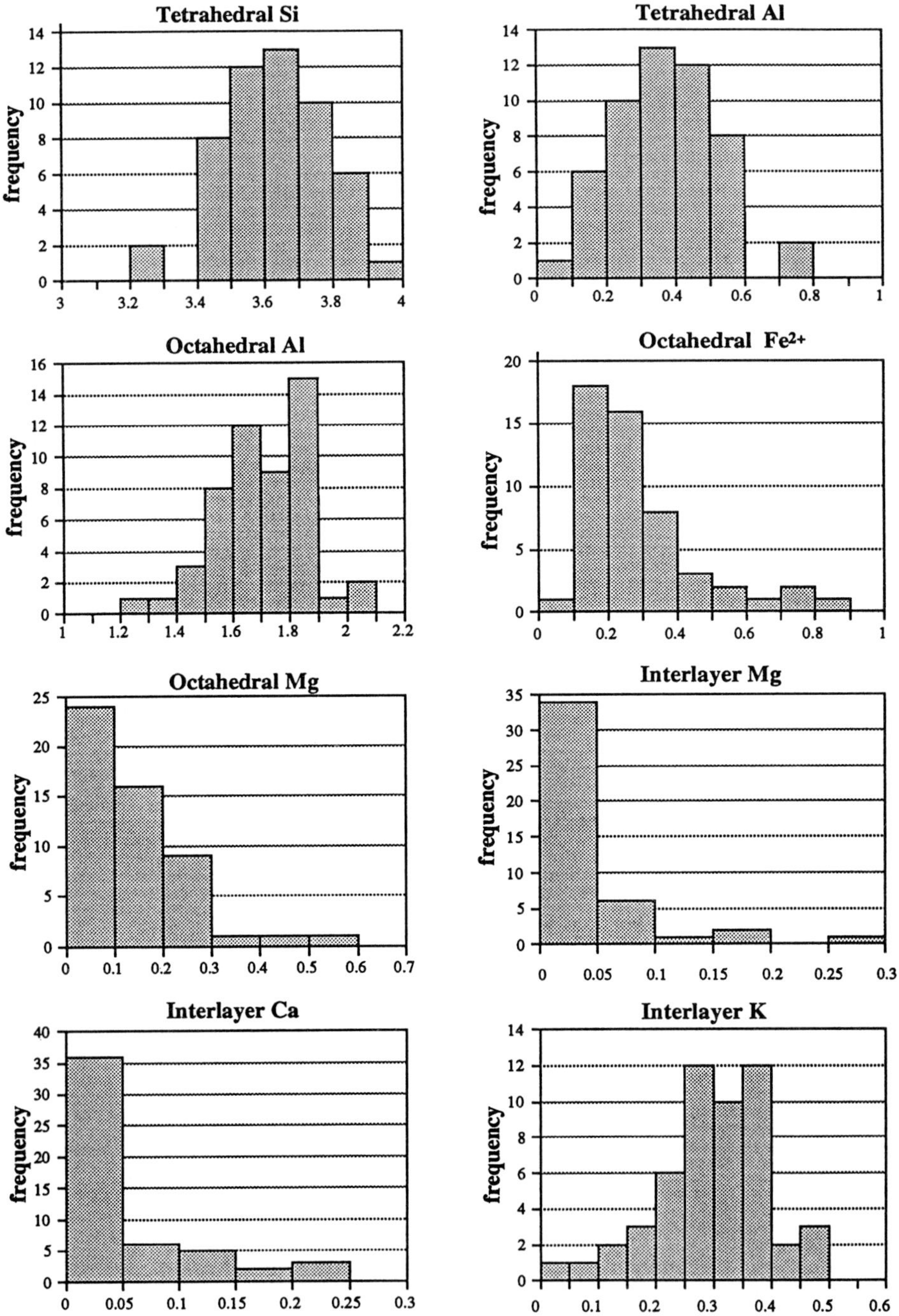

Fig. 10 - Histograms showing the distribution of cations in the structural formulae of 52 mixed layer illite-smectite particles from Marton I & II.

Diagenetic sequence of clay minerals-

Illitisation of smectites-

As mixed-layer IS presumably replaced smectite to a large extent, we looked for time-equivalent series in the Yorkshire area where this precursor could be preserved. Smectite has been described in the onshore Kimmeridgian deposits of the Boulonnais area (France), located in the nearshore vicinity of the London-Brabant Massif during late Jurassic times (Deconinck *et al.*, 1983 ; Proust *et al.*, 1993).

The part of the KCF from Marton 87, dated from the *Eudoxus* zone can thus be considered as a distal basinal equivalent of the smectite-rich "Calcaires et Argiles de Moulin, Wibert Formation", which was deposited in a more proximal, nearshore environment (Herbin and Geyssant, 1993). TEM observation of this smectite reveals a typical flaky morphology with crumpled and curled edges (Plate 2). On XRD diffractograms, d-spacing of 001 and 003 reflections for glycol-solvated samples range respectively from 16.79 to 17.10 Å and from 5.54 to 5.62 Å, suggesting a randomly interstratified structure with 70 to 100% smectite layers. The structural formulae, inferred from EDS analyses (Table 2), correspond to iron-beidellite species with an averaged K_2O content of 2.6%, reflecting the low percentage (0-30%) of illite layers. Illitisation of smectite implies a decrease in the SiO_2/Al_2O_3 ratio and an increase in the K_2O content. As expected, on a Velde ternary diagram, the ordered IS from Marton and Ebberston boreholes are located in the middle of the diagenetic trend joining illite and smectite (fig. 7). Moreover, when all the morphological and chemical steps between smectite and IS (35 to 65% of illite layers) can be demonstrated (Plate 2), the illite end-member has not been clearly identified in our samples.

Kaolinitisation of mica-

Kaolinite frequently appears as pseudomorph or overgrowth forms on more or less *in situ* weathered mica minerals (Plate 3). A diagenetic sequence is obvious on the Velde diagram between the illite and kaolinite poles (fig. 7). We attempted to estimate the degree of *in situ* alteration of micaceous phases using the excess of octahedral occupancy when all mineral formulae are calculated on an 11 oxygen atom basis. Kaolinite will give theoretically 2.29 atoms in octahedral sites (Velde and Nicot, 1985). Using this convention, it becomes possible to calculate the kaolinite content leading to an excess of occupancy. Results indicate that the transformation of mica into kaolinite can stretch as far as 75% of the support; in most cases, distinct relicts of detrital micaceous clays could still be observed (Plate 1).

Neoformed kaolinite-

In addition to very thin sub-euhedral particles of kaolinite, always bound to weathered micas and presenting the very common anhedral crystals of detrital origin, we also found isolated particles of fine-grained and well-shaped euhedral forms of this clay species (Plate 3). The origin of those pseudo-hexagonal particles remains questionable. Sedimentary kaolinites are often of disordered types and rarely show a tendency to euhedral habitus. In our opinion, the latter are thought to represent pores or coccolith infillings infilled with authigenic products, which developed in the available space. They have often been described in microtexture and fabrics studies of organic-rich sediments from the KCF (Belin and Brosse, 1992). Recent studies of late Jurassic sandstones from the North Sea (Bjørlykke and Aagaard, 1992; Burtley and McQuaker, 1992) suggested kaolinite precipitation related to detrital feldspar alteration, this contribution to the kaolinite content could not be evidenced in the samples studied here.

Controls on clay diagenesis-

Numerous mechanisms can be invoked to account for the observed clay diagenesis and organic matter maturation in the KCF. Organic geochemical and petrographical studies coupled with Rock Eval analyses indicated that the organic matter is immature to poorly mature (Tmax values ranging from 420 to 430°C; Ramanampisoa *et al.*, 1992). Productivity, redox conditions of the depositional environment and sedimentation rate appear as the main driving force for OM accumulation and preservation (Tribovillard *et al.*, 1994). Metabolisable-OM content and the presence of inorganic phases (reactive iron and carbonate contents) are the major controls on the duration and course of OM degradation and early diagenesis reactions taking place within the sulphate reduction zone, the methanogenesis zone and the thermal decarboxylation zone. Finally, the effects of burial depth can be best defined by the degree of OM maturation.

Illitisation reactions depend on factors such as burial temperature as well as porewater chemistry. Studies of offshore North Sea wells of middle Jurassic to Oligocene age showed that changes in the composition and structure of interstratified illite-smectite clays may be generated by an increasing temperature during burial and may be correlated with temperature-related changes in the organic fraction of the sediments (Pearson *et al.*, 1982). Particularly, conversion of randomly ordered IS is related to the oil window maturation stage. Nevertheless, the onshore KCF is organically immature and, according to Oschmann (1987, 1989), IS are mainly ordered with local variations in the smectite content. IS from Ebberston I (2.8% of K_2O on average) exhibit mainly randomly ordered structures whereas those from Marton I and II (3.4 to 4.1% K_2O) are typically ordered. On the other hand, time-equivalent samples from the Dorset area (Blackstone bed 42 of Cox and Gallois, 1981) show the most evolved illitisation processes (4.7% of K_2O on average; Table 2, Plate 3) with IS close to the illite end-member. Consequently, neither disparities in age, burial depth (nearly 1400 m in Marton and Ebberston; Williams, 1986) nor lithology (carbonate content) are satisfactory explanations for this structural trend of random-ordering. Therefore, an additional control, besides the temperature gradient, has to be invoked to account for the degree of illitisation. As discussed by Scotchman (1987), and in agreement with this author, the most likely additional mechanism is a chemical one, involving porewater chemistry and/or K-supply control. During OM diagenesis within sediments, reactions of dissolution and precipitation of carbonates and Al-silicates are pH-dependent (Curtis, 1987). Microbially-induced OM degradation will add H^+ protons into the milieu; lowered pH can destabilise detrital Al-silicates, *e.g.* micas and, possibly, detrital smectites. Alteration of the latter two minerals will provide Al and Si for kaolinite authigenesis in slightly acidic environments and, K, Fe and Mg are readily extensively leached. Oxidation and methanogenesis zones are more propitious to these dissolution-precipitation reactions than the sulphate reduction zone. However, another important chemical reaction occurs in the sulphate reduction zone where OM, acting as a reducing agent for Mn and Fe oxihydroxides, causes pH raising. In this case, an increase in alkaline porewater conditions can favour illitisation processes as well as iron-carbonate and dolomite precipitation (Irwin, 1977; 1980). These considerations suggest that the balance between OM content and iron availibility and, on the other hand, pore-fluid composition has a major influence on illitisation processes.

Mass-balance calculation-

We assume that diagenetic processes occurred in a closed system without removal or supply of ions that implies a closed relationship between kaolinitisation and illitisation reactions. In other words, leaching of micas will provide Si and Al atoms for kaolinite precipitation and K atoms for illite layers in mixed-layers IS. In addition, illitisation of smectite, through IS formation, will induce an excess of Si and quartz authigenesis.

Taking into account for each sample:
- the chemical composition (SEM-EDS) of the clay fraction and that (TEM-EDS) of the clay minerals (kaolinite, mica, IS and , as a precursor, smectite),
- the percentage, estimated after XRD and FTIR analyses, of clay mineral species,
we can determine the required amount of leached micas for supplying K atoms to illite layers in smectite and, consequently, estimate the amount of authigenic kaolinite.

The main conclusions of this theoretical calculation of K exchange between mica and smectite minerals are (Table 3):

- averaged percentage of authigenic kaolinite compared to the total kaolinite content varies between 10 and 20%
- variations in authigenic kaolinite from one cycle to the other lead to different degrees of interstratification (40 to 60% of illite layer) in mixed-layer IS, in good agreement with observed values (XRD and TEM/EDS analyses; Table 2) ;
- samples from Marton I show the highest degree of interstratification, despite their low OM content, as discussed above, other controlling factors, such as lithology, extension of the sulphate reduction zone and values of the degree of pyritisation (DOP defined by Berner, 1970; see Tribovillard *et al.*, 1994 and Lallier-Vergès *et al.*, this volume) have to be considered;
- the release of Si, Fe, Mg atoms during illitisation of smectite could explain both pre-cipitation of diagenetic quartz and ankerite nodule formation as described by Irwin (1977, 1980);
- perhaps the most important result drawn from mass-balance calculation is that, through each organic cycle, the amount of authigenic kaolinite remains roughly constant. Thus, by subtracting the effects of diagenesis, we could infer the primary flux of detrital clay minerals (kaolinite, mica, smectite; Fig 3, 4 and 5) and tentatively interpret their variations in terms of erosion/alteration inputs.

Palaeoclimatic significance-

Clay mineral fluctuations on high-frequency scale-

A similar clay mineral pattern is observed (fig. 3, 4 and 5) through each third-order cycle studied: kaolinite abundance relative to IS and, to a minor degree, to mica reaches its maximum near the peak of the TOC curve. Kaolinite and mica or IS opposite distribution are corroborated by the negative correlation of the K/Al ratio with kaolinite (fig. 11). From the above, changes in kaolinite abundance alone do not reflect diagenetic signature and, after subtracting diagenetic effects, calculated primary-kaolinite abundance also increases, relative to mica and inferred primary-smectite abundance (fig. 3, 4 and 5).

Many authors claim that grain-size sorting plays a prominent role in determining the differential settling of clay minerals in marine environments, with kaolinite and illite tending to be deposited in nearshore, shallow-water, settings (Chamley, 1989). Our preliminary results indicate that, for a time-equivalent period (the *Eudoxus* or *Wheatleyensis* zone), the kaolinite/IS ratio displays a similar spatial pattern throughout

MARTON I

Depth (m)	TOC (%)	1	2	3	4
-128.14	2.54	6	7.7	2.4	62
-128.34	2.89	8.1	10.7	3.2	58.9
-128.4	3,00	6.6	8.5	2.7	59.1
-128.52	3.55	7.2	9.5	2.9	64.6
-128.68	5.64	6.3	8.4	2.6	69.9
-128.77	7.99	5.6	7.3	2.3	71
-128.86	6.71	6	7.2	2.3	65.9
-128.91	4.91	6.9	8.9	2.8	58.8
-128.97	2.34	7.8	10.1	3.1	55
-129.05	1.9	8.1	10.5	3.2	53
-129.25	1.88	9.9	13.4	3.9	57.5
-129.27	2.21	7.5	9.7	3	56.2
Mean		**7.2**	**9.3**	**2.9**	**61**

MARTON II

Depth (m)	TOC (%)	1	2	3	4
-121.6	4.73	5.5	6.7	2.9	43.9
-121.57	5.53	4.7	5.8	2.5	44.1
-121.54	9.19	5.3	6.6	2.8	49.7
-121.5	10.15	6.3	8.1	3.3	57.8
-121.47	14.79	5.4	7	2.9	63.9
-121.44	20.04	4.1	5.1	2.2	55.5
-121.42	28.94	3.6	4.4	2	55.5
-121.4	27.87	3.6	4.4	2	54.8
-121.37	17.5	4.9	6.2	2.6	57.9
-121.35	13.66	5.6	7.4	3	65.1
-121.3	16.86	5	6.3	2.6	58.2
-121.28	9.18	5.9	7.7	3.1	59.6
-121.27	13.08	6.2	8.3	3.2	68.5
-121.24	15.06	5.3	6.9	2.8	64.7
-121.21	9.49	5.2	6.7	2.8	57.9
-121.17	7.39	5.7	7.4	3	63.9
-121.1	8.14	6.4	8.5	3.4	61.4
-121.07	17.02	7.9	10.7	4.1	58.8
-121.01	9.28	7.9	11.1	4	67.6
Mean		**5.5**	**7.1**	**2.9**	**58.4**

EBBERSTON I

Depth (m)	TOC (%)	1	2	3	4
-69.2	3.04	4.2	6.1	5.5	33.7
-69.35	3.04	3.8	5.5	5.1	34.3
-69.67	5.3	3	4.3	4.1	34.8
-69.99	11.86	3.6	5.4	4.9	41.4
-70.02	3.69	3	4.2	4.1	33.2
-70.06	17.04	2.4	3.6	3.4	42.5
-70.16	8.96	3.3	5	4.5	46.9
-70.23	13.38	2.5	3.7	3.6	41.2
-70.25	18.21	2.8	4.2	3.9	49.8
-70.32	6.79	2.7	3.9	3.8	36.8
-70.38	4.74	4.4	6.9	5.7	49.1
-70.58	3.55	2.6	3.7	3.7	29.3
-70.81	6.23	3.5	5.3	4.8	44
Mean		**3.2**	**4.8**	**4.4**	**39.8**

Table 3 - Prediction of diagenetic budget from mass-balance calculation. Percentage in the actual clay fraction of 1) diagenetic kaolinite replacing mica 2) replaced mica, 3) neoformed quartz, 4) illite-layers in mixed-layer illite-smectite.

the Cleveland Basin, from the basin setting (Marton) to the platform area (Flixton). The same conclusion can be drawn from comparisons between two neighbouring basins (Dorset and Yorkshire) and with shelf, organic-poor, deposits (Boulonnais area). This means that the relative abundance of these pedogenetic clay species cannot be related to differential settling patterns. This is certainly not true for illite distribution: together with a climatic influence, the main determining parameters for illite abundance could be the petrographical composition of source area (London-Brabant Massif and/or Grampian Massif, Mid North-Sea High) and, more probably, differential settling between distal basins (Dorset and Yorkshire) and proximal areas (Boulonnais).

Eustatic sea-level changes controlled either by climate or by tectonics are often invoked as an explanation for the succession of clay mineral assemblages. As a consequence of tectonic rejuvenation leading to shoreline displacement, the relative abundance of kaolinite, compared to that of illite or smectite, can be used to infer shallowing/regressive and deepening/transgressive episodes (identified by an increased supply of kaolinite and/or illite, and an increase in smectite supply, respectively; Chamley, 1989; Proust and Deconinck, 1993). However, possible tectonic/eustatic sea-level fluctuations occur on a much coarser scale than that of clay-mineral variations within a third-order cycle, whereas, as mentioned above, the kaolinite/smectite ratio appears as unrelated to differential settling and, consequently, to the shoreline location.

Kaolinite abundance as a palaeoclimatic indicator

It is thought that a sudden major climatic change occurred during Kimmeridgian times in north-western Europe (Wignall and Ruffel, 1991). Various sedimentological, palaeoecological and geochemical characteristics of marine deposits changed during the interval of time comprised between the lower part of the *Huddlestoni* zone to the mid *Pectinatus* zone, as recorded in the upper part of the KCF in southern England. Notably, a decline in kaolinite and illite abundance and, correlatively, an increase in smectite abundance has been inferred as a climatic indicator of humid to semi-humid conditions. Therefore, humid and warm conditions must have prevailed during the *Eudoxus* and *Wheatleyensis* zones. Nevertheless, as outlined by Oschmann (1988), short-term climatic oscillations (5.10^3 to 15.10^3 years) offer the most suitable interpretation for the third-order cyclicity.

Two arguments could militate in favour of a climatic influence:
- kaolinite abundance, when compared to other clay minerals, displays an almost linear trend *vs.* TOC values up to TOC = 7% and then becomes progressively unrelated (fig. 12). Previous studies (Huc *et al.*, 1992; Ramanampisoa *et al*, 1992;

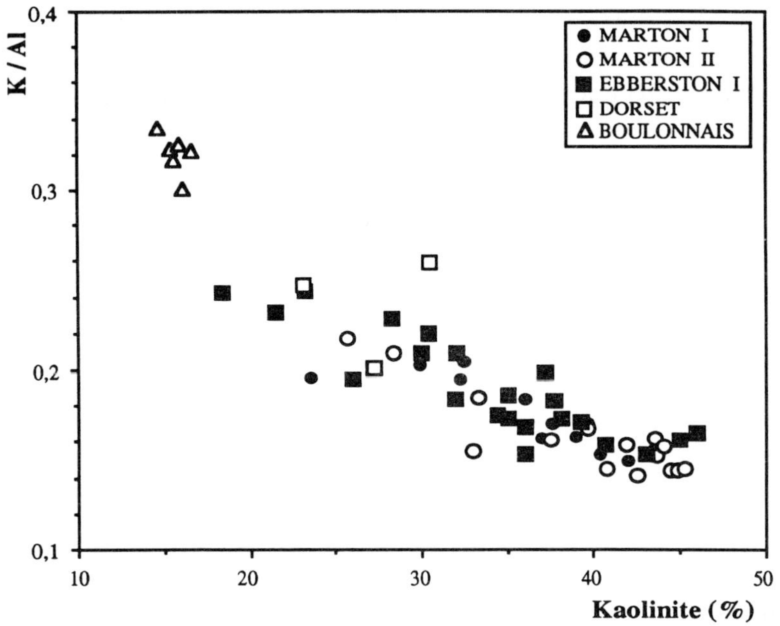

Fig. 11 - Relationship between the chemical K/Al ratio of sediments and the abundance of kaolinite.

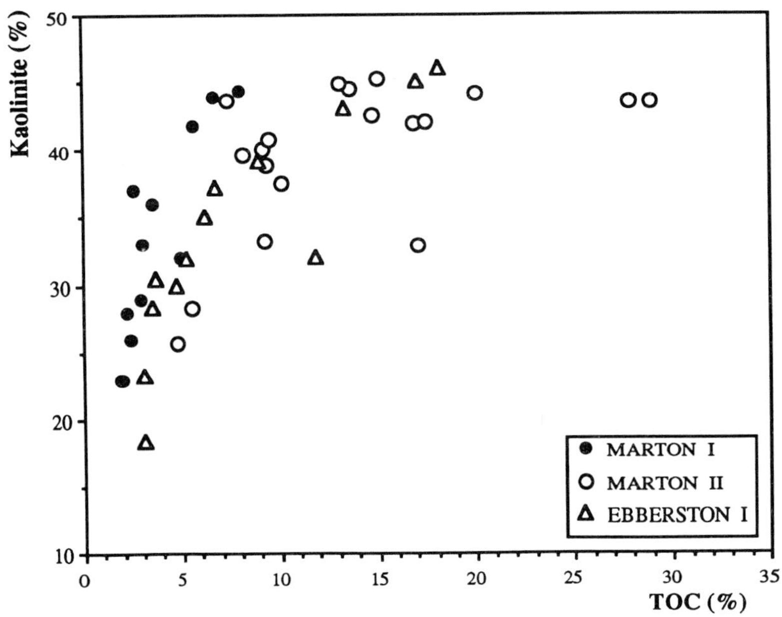

Fig. 12 - Relationship between the organic content of the sediment (TOC) and the abundance of kaolinite.

Lallier-Vergès *et al.*, 1993, Disnar and Ramanampisoa, this volume) have also shown that, below 6% TOC, organic geochemical parameters (*e.g.* Hydrogen Index or HI, molecular biomarkers) indicated variations both in the quality and in the quantity of sedimentary OM. Beyond 6% TOC, the OM nature remains constant. Compositional changes in the terrestrial flux (leading to increased sedimentation rates) could favour better preservation conditions for OM. However, through each cycle, the quartz/clay abundance ratio remains almost constant and the Si/Al ratio is not an indicator of grain size variation, being only monitored by the kaolinite content. Thus, on a third-order scale, OM productivity could be coupled, at least until TOC reaches 6%, to nutrients which would be subordinated to kaolinite inputs.

The values of the Th/K ratio from sedimentary rocks can be used in different ways. Covariant changes in Th and K content essentially reflect the proportion of the clay fraction in sediments. Changes in the Th/K ratio possibly represent variations in clay mineralogy, therefore such changes may be related to proximal/distal oscillations (prograding/retrograding systems of sequence stratigraphy) or climatic change records or both (Myers and Wignall, 1987). The data from this study show Th/K ratio values changing from one cycle to the other. However, through each individual cycle, Th/K values show an obvious relation with the values of the K/Al and Fe/Al ratios (fig. 3, 4 and 5) , and possibly, with kaolinite abundance (fig. 13). The same trend was observed by Van Buchem *et al.* (1992) for the lower Lias of Yorkshire. Reciprocally, different Th/K values for the same kaolinite content mean that the Th content does not depend only on clay mineralogy. As a matter of fact, preliminary results of sequential chemical extraction procedure (Tribovillard *et al*, 1994) indicate that, for Ebberston I, more than 50% of Th content is bound to iron oxihydroxides and that a substantial amount of thorium is incorporated to Th-rich accessory minerals such as monazite (work in progress).

To summarise, iron, thorium and kaolinite are probably good indicators for the nature and for the source areas of mobilised soils. Chemical and mineralogical profiles indicate conditions were more humid and warmer during the time of deposition of the most OM-rich part of elementary cycles. Co-occurrence of the signals of increasing weathering on land and of enhanced productivity in the basin could be inferred from the nutrient influx.

Conclusions-

For the interpretation of clay mineral assemblages in fine-grained sediments from basinal settings, two main factors (on third-order scale) must be considered:
- post-depositional diagenetic changes affecting the detrital supply (a),
- relationship between climatic parameters and on-land clay mineral formation (b).

Tectonic control and grain size-induced differential settling did not play a prominent role in distal environment for clay mineral distribution (concerning high-frequency cycles).

(a) No major diagenetic alteration, able to damage extensively the environmental message, could be evidenced. Authigenic kaolinite and micas replaced *in situ* do not exceed 20% of the clay fraction and this percentage remained almost constant throughout whole sections of several TOC cycles. On the other hand, progressive transformation of smectite into mixed-layer IS never reached the pure illite end-member pole.

(b) The distribution of major and trace elements as well as changes in clay mineral assemblages are well correlated with TOC variations suggesting a climatic control on third-order cycles. Our next aim, concerning the basinal setting, is to examine the way regressive and transgressive phases are recorded at lower cycle frequency.

Acknowledgements- We thank Pierre Tremblay and Nicole Moureau for the analytical support. This work benefited greatly from the help of the Institut Français du Pétrole.

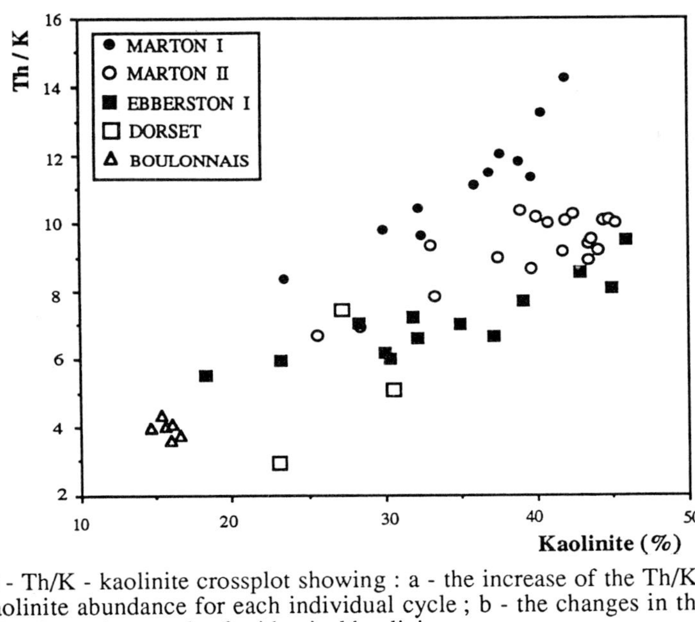

Fig. 13 - Th/K - kaolinite crossplot showing : a - the increase of the Th/K ratio values with kaolinite abundance for each individual cycle ; b - the changes in the Th/K ratio values for the various cycles for identical kaolinite contents.

References-

Belin S. and Brosse E. (1992) Petrographical and geochemical study of a Kimmeridgian organic sequence (Yorskhire area, UK). *Rev. Inst. Fr. Pétr.*, **47**, 711-725.

Berner R.A. (1970) Sedimentary pyrite formation. *Am. J. Sci.*, **268**, 1-23.

Bertrand P and Lallier-Vergès E. (1993) Past sedimentary organic matter accumulation and degradation controlled by productivity. *Nature*, **364**, 786-788.

Bjørlykke, K. and Aagaard P. (1992) Clay minerals in North Sea sandstones. *In:* Origin, Diagenesis and Petrophysics of Clay Minerals in Sandstones. SEPM Special Publication N° 47, pp. 65-80.

Burley S.D. and MACQuaker J.H.S (1992) Authigenic clays, diagenetic sequences and conceptual diagenetic models in contrasting basin-margin and basin-center North Sea Jurassic sandstones and mudstones. *In* Origin, Diagenesis and Petrophysics of Clay Minerals in Sandstones. SEPM Special Publication N° 47, pp. 81-110.

Chamley H. (1989). Clay Sedimentology. Heidelberg, Berlin, New york: Springer-Verlag. 623 p.

Cox B.M.and Gallois R.W.(1981) The stratigraphy of the Kimmeridge Clay of the Dorset type area and its correlation with some other Kimmeridgian sequences. *Rep. Inst. Geol. Sci. London*, **80** (4), 1-44.

Curtis C (1987). Mineralogical consequences of organic matter degradation in sediments: Inorganic / Organic diagenesis. *In* LEGGETT J. K and ZUFFA G.G. (Eds), Marine Clastic Sedimentology, Graham and Trotman, London, p 108-123.

Deconinck J.F., Chamley H., Debrabant P. and Colbeaux J.P.(1983). Le Boulonnais au Jurassique supérieur: données de la minéralogie des argiles et de la géochimie. *Ann. Soc. Géol. Nord*, CII. p. 145-152.

Hallam A., Grose J.A. and Ruffell A.H. (1991). Palaeoclimatic significance of changes in clay mineralogy across the Jurassic-Cretaceous boundary in England and France. *Palaeogeogr. Palaeoclimatol. Palaeoecolol.,* **81**, 173-187.

Herbin J.P., Geyssant J.R., Müller C., Mélières F., le groupe YORKIM. and Penn, I.E. (1991). Hétérogénéité quantitative et qualitative de la matière organique dans les argiles du Kimméridgien du val de Pickering (Yorkshire, UK). Cadre sédimentologique et stratigraphique. *Rev. Inst. Fr. Pétr.*, **46** (6), 1-39.

Herbin J.P. and Geyssant J.R.(1993). "Ceintures organiques" au Kimméridgien / Tithonien en Angleterre (Yorkshire, Dorset) et en France (Boulonnais). *C.R. Acad. Sci. Paris*, **317**, 1309-1316.

Herbin J.P., Geyssant J.R., Mélières F., Müller C., Penn I.E and YORKIM group (1993) Variation of the distribution of organic matter within a transgressive system tract: Kimmeridge Clay (Jurassic), England. *In* AAPG - Studies in Geology. Petroleum Source Rocks in a Sequence Stratigraphic Framework, B. Katz and L. Pratt. (éds) pp. 67-99.

Huc A.Y., Lallier-Vergès E., Bertrand P., Carpentier B. and Hollander D.J. (1992). Organic matter response to change of depositional environment in Kimmeridgian shales, Dorset, U.K. *In:* Organic matter productivity, accumulation and preservation in recent sediments, J. Whelan and J. Farrington (Eds), Columbia University Press, New York pp. 469-486.

Irwin H.and Curtis C. (1977). Isotopic evidence for source of diagenetic carbonates formed during burial of organic-rich sediments. Nature, **269**, 209-213.

Irwin H. (1980). Early diagenetic carbonate precipitation and pore fluid migration in the Kimmeridge Clay of Dorset, England. *Sedimentology,* **27**, 577-591.

Lallier-Vergès E., Bertrand P., Huc A.Y., Bückel D. and Tremblay P. (1993). Control of the preservation of organic matter by productivity and sulphate reduction in Kimmeridgian shales from Dorset (U.K). *Mar. Petrol. Geol.*, **10**, 598-605.

Myers K.J. and Wignall P.G. (1987) Understanding Jurassic organic-rich mudrocks — New concepts using Gamma-ray Spectrometry and Palaeoecology: Examples from the Kimmeridge Clay of Dorset and the Jet rock of Yorkshire. *In* Marine Clastic Sedimentology, LEGGETT J. K and ZUFFA G.G. (eds.) Graham and Trotman, London, p 172-189.

Oschmann W. (1988). Kimmeridge Clay sedimentation — A new cyclic model. *Palaeogeogr. Palaeoclimatol. Palaeoecol.*, **65**, 217-251.

Pearson M.J., Watkins D. and Small J.S. (1982) Clay diagenesis and organic maturation in Northern North Sea sediments. *In*. "Int. clay. Conf., Bologna - Pavia, 1981 VAN OLPHEN H. and VENIALE F. (eds.). Elsevier, Developments in Sedimentology, 35, p 665-675.

Pradier B.and Bertrand P. (1992) Etude à haute résolution d'un cycle du carbone organique des argiles du Kimmeridgien du Yorkshire (G.B): relations entre composition pétrographique du contenu organique observé *in situ*, teneur en carbone organique et qualité pétroligène. *C.R. Acad. Sci. Paris*, **315** (2), 187-192.

Proust J.N., Deconinck J.F., Geyssant J.R., Herbin J.P. and Vidier J.P. (1993) Nouvelles données sédimentologiques dans le Kimméridgien et le Tithonien du Boulonnais (France). *C.R. Acad. Sci. Paris*, **316**, 363-369.

Ramanampisoa L., Bertrand P., Disnar J.R., Lallier-Vergès E., Pradier B. and Tribovillard N.P. (1992) Etude à haute résolution d'un cycle du carbone organique des argiles du Kimméridgien du Yorkshire (G.B.): résultats préliminaires de géochimie et de pétrographie organique. *C.R. Acad. Sci. Paris*, **314** (2), 1493-1498.

Reynolds R.C. and Hower J. (1970) The nature of interlayering in mixed-layer illite-montmorillonites. *Clays and Clay Minerals*, **18.** 25-36.

Scotchman I.C. (1987) Clay diagenesis in the Kimmeridge Clay Formation, onshore UK, and its relation to organic maturation. *Miner. Mag.*, **51**, 535-551.

Scotchman I.C. (1989) Diagenesis of the Kimmeridge Clay Formation, onshore U.K. *J. Geol. Soc. London*, **146**, 285-303.

Tribovillard N.P., Desprairies A., Bertrand P., Lallier-Vergès E., Disnar J.R. and Pradier B. (1992) Etude à haute résoulution d'un cycle du carbone organique de roches kimméridgiennes du Yorkshire (Grande-Bretagne): minéralogie et géochimie (résultats préliminaires). *C.R. Acad. Sci. Paris*, t. 314, Série II, 923-930.

Tribovillard N.P., Desprairies A., Lallier-Vergès E., Bertrand P., Moureau N., Ramdani A. and Ramanampisoa L. (1994) Geochemical study of organic-matter rich cycles from the Kimmeridge Clay Formation of Yorkshire (UK): Productivity versus anoxia. *Palaeogeogr. Palaeoclimatol. Palaeoecol.,* **108**, 165-181.

Van Buchem F.S.P., Melnyk, D.H., and McCave, I.N. (1992) Chemical cyclicity and correlation of Lower Lias mudstones using gamma ray logs, Yorkshire, UK. *J Geol. Soc., London,* **149**, 991-1002.

Velde B. (1977) Clays and Clay minerals in natural and synthetic systems. Amsterdam, Elsevier. 218 p.

Velde B. and Nicot E. (1985) Diagenetic clay mineral composition as a function of pressure, temperature and chemical activity. *J. Sed. Petrol*., **55**, 541-547.

Wignall P.B. and Ruffell A.H. (1990) The influence of a sudden climatic change on marine deposition in the Kimmeridgian of northwest Europe. *J. Geol. Soc. London,* **147**, 365-371.

Williams P.F.V. (1986) Petroleum geochemistry of the Kimmeridge Clay of onshore southern and eastern England. *Mar. Petrol. Geol.,* **3**, 258-281.

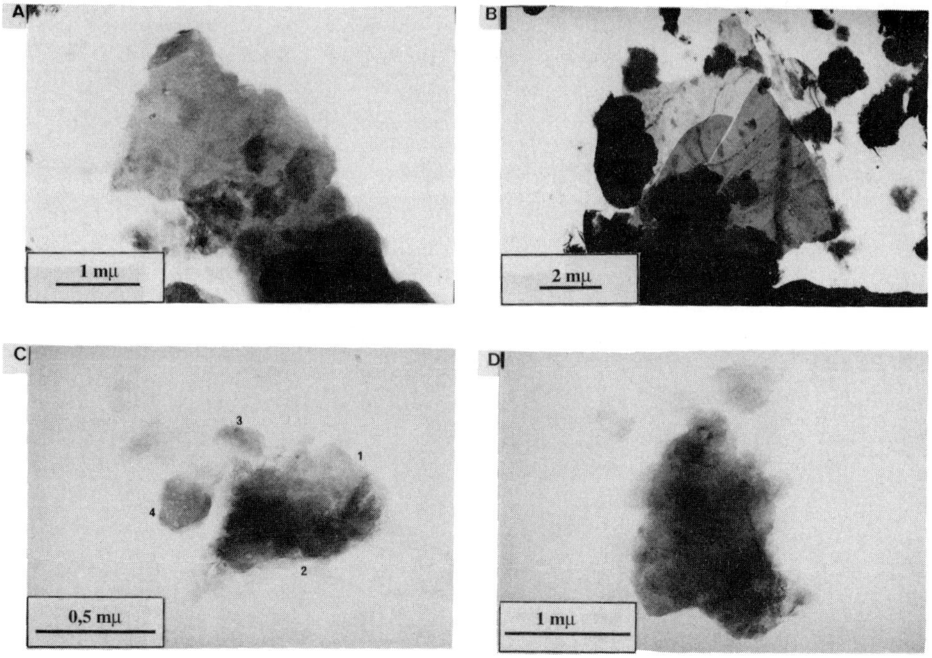

Plate 1.

Electron micrographs showing the morphology of detrital and diagenetic clay minerals occurring in Marton I and II cycles.

A- Anhedral particle of detrital Kaolinite.

B- Euhedral crystal of detrital Mica clay mineral.

C- Mica (1) particle overgrown with diagenetic Kaolinite (2) and euhedral crystals of authigenic Kaolinite (3-4).

D- Diagenetic Kaolinite almost entirely replacing detrital Mica.

Samples. A. Marton II - 121.60 m. TOC 4.7 %.
B. Marton II - 121.30 m. TOC 16.9 %.
C. D. Marton I - 128.14 m. TOC 2.5 %.

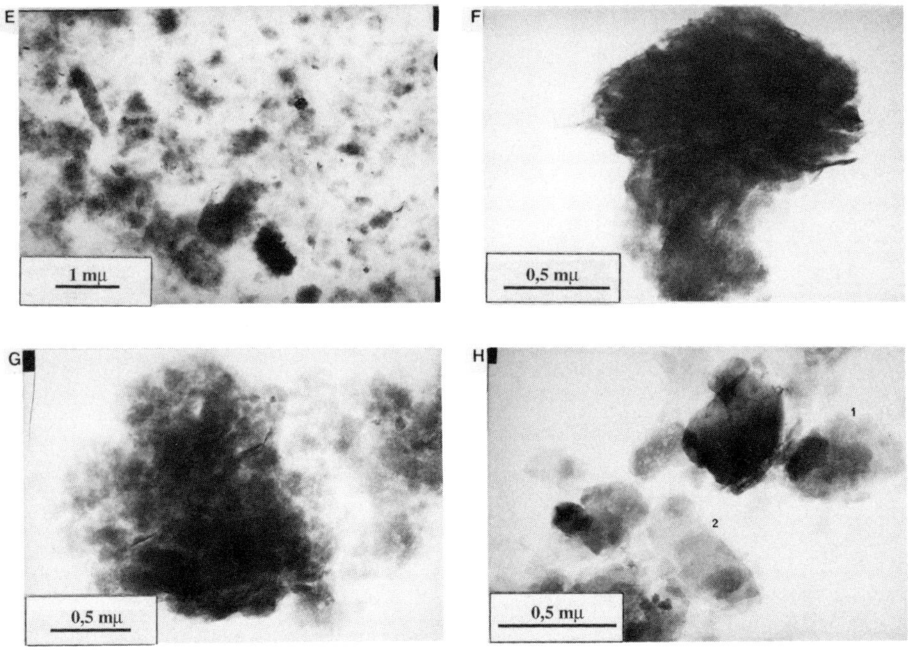

Plate 2.

Electron micrographs illustrating diagenetic pathway of illitisation in Marton and Ebberston cycles.

E- Thin flakes of smectite from Boulonnais (France), stated as equivalent precursor for mixed layer illite-smectite (IS) of Cleveland basin

F.G- Randomly ordered IS (35 to 45 % of illite layers) exibiting progressive growth of thin platelets at the curled edges of flaky aggregates.

H- Ordered IS with 50 % (1) to 65 % (2) of illite layers, showing development of lath-like particles with sub-euhedral shape.

Samples. E. Boulonnais- "Calcaires et Argiles de Moulin Wibert"- *Eudoxus* zone.
F. Marton II - 121.40 m. TOC 27.9 %.
G. Ebberston I - 70.25 m. TOC 18.2 %.
H. Marton II - 121.24 m. TOC 15.1 %.

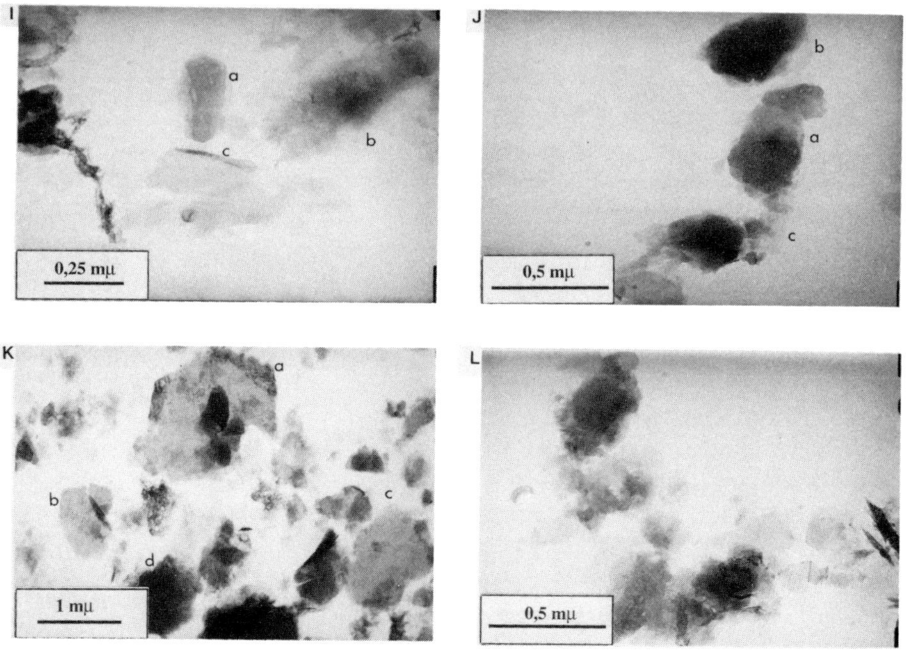

Plate 3.

Electron micrographs of clay mineral assemblages found in the Kimmeridge Clay Formation.

I - Well shaped authigenic Kaolinite (a) associated to mixed layer IS showing relation of their morphology with illitisation reaction: (b) and (c) are IS with respectively 35 and 60 % of illite layers.

J - Thin crystals of diagenetic Kaolinite (a) entirely replacing Mica and two specimens (b-c) of mixed layer IS containing betwen 50 and 60 % of illite layers.

K- Representative assemblage in the KCF of detrital (a- mica b- kaolinite) and diagenetic (c- IS with 45 % of illite layers; d kaolinised mica) clay minerals.

L- Sub-euhedral particles of mixed layer IS (85 % of illite layers) illustrating the most advanced stage, achieved in Dorset area, of conversion of smectite to illite.

Samples. I. Marton II. 121.47 m. TOC 14.8 %.
J. Marton I. 129.25 m. TOC 1.9 %.
K. Marton II. 121.57 m. TOC 5.5 %.
L. Dorset. "blackstone" bed N° 42. TOC 48 %.

THE RECENT MAAR SEDIMENTS
(LAC DU BOUCHET, FRANCE)

Geochemical study of the Lac du Bouchet, Haute-Loire, France
Part I : water balance and biogeochemical implications

Eric Viollier[1], *Patrick Albéric*[2], *Marc Evrard*[1], *Didier Jézéquel*[1], *Dominique Lavergne*[1], *Gil Michard*[1], *Monique Pèpe*[1], *Gérard Sarazin*[1] *and Pierpaolo Zuddas*[1]

1) Laboratoire de Géochimie des Eaux, Université Paris VII, case postale 7052, F-75251 Paris cedex 5
2) Université d'Orléans, URA 724 du CNRS, Dépt. des Sciences de la Terre, F-45067 Orléans cedex

Key words- *water balance, water column, biogeochemical processes, oligotrophic, trace elements, iron cycle.*

Abstract- Chemical survey of the Lac du Bouchet crater lake and analyses of close springs and rainfall allowed to understand hydrological functioning. Without river input or output, lake is mainly fed by rainfall. Although underlacustrine spring water feeding remains possible, its contribution needs not be taken into account. Annual evapotranspiration at the lake surface has been found equal to 0.7 time rain inputs. Furthermore, substantial seepage (48% of total inputs) occurs through sediments leading to a chemical leak. Therefore, limited nutrients injection in the water column defines the oligotrophic status of the lake. Hypolimnion organic matter biodegradation takes place from May to November essentially by mean of dissolved O_2 while epilimnion is always very close to equilibrium with atmospheric O_2 and CO_2. Associated with temporary anoxic conditions taking place in November just over the bottom, Co, As, Mo, V, Ce, Pb, and Al are closely related to Fe diffusing from interstitial water.

Introduction-

Continental closed basins such as lakes are of great interest for the study of early diagenesis, since their small dimensions allow better definition of geochemical systems than for oceans. The Lac du Bouchet is a maar lake, furthermore specifically interesting because of : 1) a continuous slow sedimentation during the last 0.35 My (Truze 1990) and a well known basaltic environment (Mergoil, 1987; Teulade *et al.,* 1991), 2) low anthropogenic influences. In order to understand organic matter production or biodegradation sequences, it is necessary to characterise both water column and interstitial waters of shallow sediments. In this paper, results from an annual survey concerning dissolved species in lake water, close springs and rainfalls are introduced and discussed to determine : 1) overall hydrological functioning, 2) the main biogeochemical processes in each season, 3) trace element distribution related to the redox interface position. Results about interstitial waters are presented in Jézéquel *et al.* 1994.

Study area-

The Lac du Bouchet is located 15 km south of Le Puy-en-Velay at an elevation of 1205m. Its shape is circular with a mean diameter of 750 m and a maximum depth of 28 m. The lake covers an area of 0.44 km^2 and its drainage basin (0.97 km^2) is largely forested. Water fills a crater formed by a phreatomagmatic explosion about 0.7 My ago (Teulade *et al.*, 1991). All the water entering the lake is from rainfall, surface runoff and possibly ground waters (there is no visible river input or output above lake surface). Springs were sampled within a radius of 5 km around and rainfall 3 km far from the lake at Cayres meteorological station.

Methods-

Fourteen vertical profiles were obtained between 3 April 1992 and 4 November 1993 at 3 midlake stations (fig. 1) in order to confirm horizontal homogeneity. Water samples were collected at 2 or 3 m intervals with a 5 litre Van Dorn PVC hydrobottle. Temperature, pH, and dissolved oxygen measurements were made in situ with 2 ®WTW probes. Analytical methods for major and minor elements and their analytical accuracy are summarised in Table 1. Water samples were filtered by a cellulose nitrate 0.45 µm membrane with nitrogen over pressure. Samples were then acidified with ®Suprapur HNO_3 to pH 2 (except for anions measurement) and stored in 125 ml polypropylene bottles. Samples for organic carbon were filtered by a precombusted glass fibre filter (®Whatman GF/F). Filter were used for particulate organic carbon (POC) determination. Dissolved organic carbon (DOC) was determined on water samples filtered by membrane or glass fibre filters. Nutrient analyses were usually analysed within 5 days of collection since fieldwork analyses did not show any significant differences. All samples were analysed for several trace elements in a semi-quantitative way by ICP-MS (Inductively Coupled Plasma - Mass Spectrometry). The water column sampled in November 1993 was obtained with a lab-made *in situ* filtration apparatus and analysed quantitatively in ICP-MS.

Results-

Temperature, oxygen and pH-

The physical and chemical properties are presented in time-depth diagrams to provide an overview of the yearly cycle. The Lac du Bouchet is dimictic (Casta 1991). Thermal

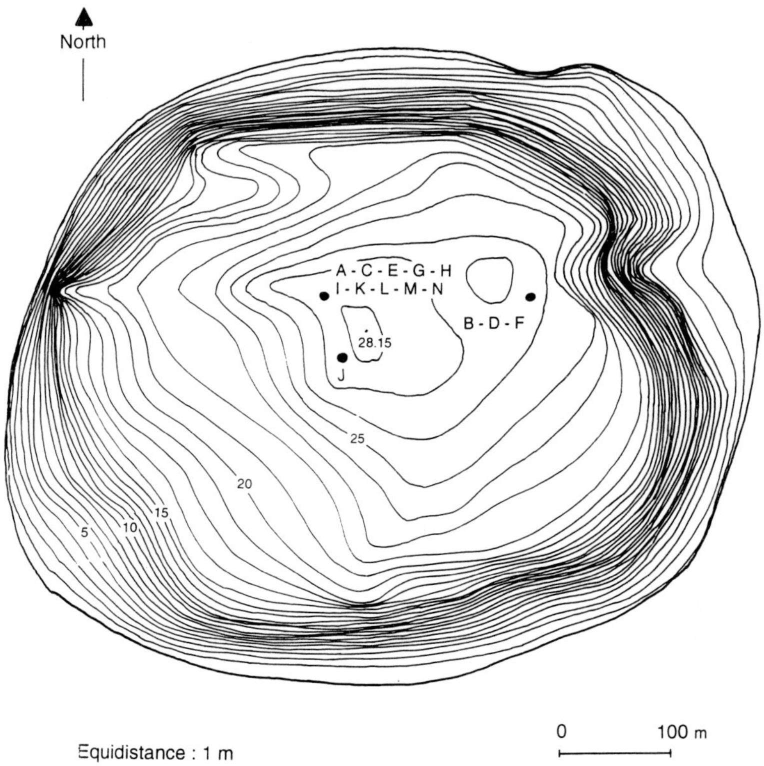

Fig. 1 - Bathymetric chart of Bouchet Lake
(after Reille and Beaulieu. 1988)

stratification begins in May and lasts until November (fig. 2). Cooler air temperature and wind turbulence cause the mixed layer to deepen during October and November and by early January the lake is isothermal. The distribution of oxygen shows distinct features (fig. 3). In the epilimnion, dissolved oxygen is close to theoretical saturation.

Table 1: Analytical methods for major and minor elements and their analytical accuracy.
FAAS : Flame Atomic Absorption Spectrometry
FAES : Flame Atomic Emission Spectrometry
GFAAS : Graphite Furnace Absorption Spectrometry
HPLC : High Pressure Liquid Chromatography
FIA : Flow Injection Analysis

Component	**Technique**	**Accuracy (%)**	**References**
pH	RWTW probe	± 0.01 pH units	
Dissolved O_2	RWTW probe	± 10	
DOC	Catalytic oxidation NiO, 850 °C	± 10	
POC	Combustion O_2, 1000 °C	± 10	
Alkalinity	Gran titration	± 3	Gran, 1952
Sodium	FAAS	± 3	
Magnesium	FAAS	± 3	
Calcium	FAAS	± 5	
	Colorimetric	± 3	Milligan *et al.* 1971
Potassium	FAES	± 3	
Iron	GFAAS	± 5	
Manganese	GFAAS	± 5	
Strontium	GFAAS	± 5	
Chloride	HPLC	± 2	
Sulphate	HPLC	± 3	
Nitrate > 5 µM	HPLC	± 5	
< 5 µM	Colorimetric FIA	± 3	Whitledge *et al.* 1981
Dissolved silica	Colorimetric FIA	± 3	Truesdale *et al.* 1975
Ammonium	Colorimetric FIA	± 3	U.S.E.P.A. 1984
Orthophosphate	Colorimetric FIA	± 5	Whitledge *et al.* 1981

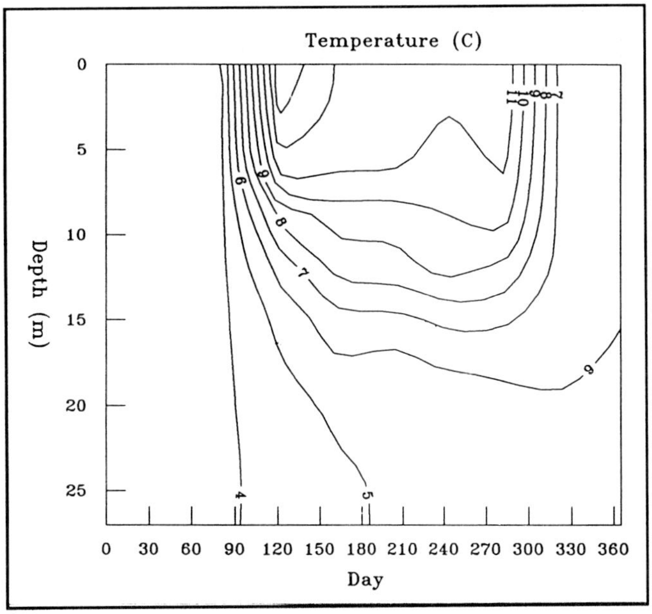

Figure 2 - Temperature isocontours in a time-depth graph

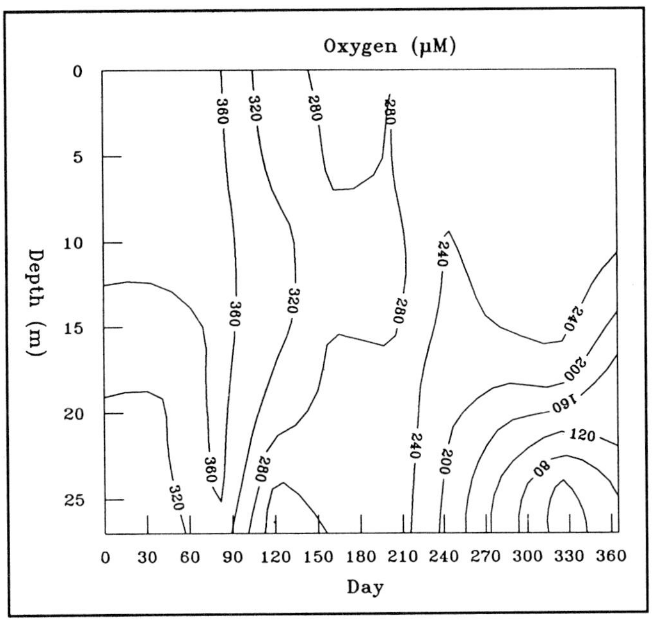

Figure 3 - Oxygen isocontours in a time-depth graph

Undersaturation occurs in autumn when surface water can be mixed with depleted deep water. Concentrations in the hypolimnion begins to decline in May to become anoxic in November a few metres above the sediment-water interface. Incorporation of : 1) autochthonous and allochthonous organic matter from the epilimnion, 2) reduced compounds diffusing from interstitial waters, in a small residual hypolimnion are responsible for consumption of oxygen during this period. The values of pH range from 6.18 to 8.53 through the year (fig. 4). The highest values occur in May and are a result of organic matter production during the spring bloom (Eckartz-Nolden and Nolden, 1991). Phyto- and zooplankton biodegradation produces acidification below the chemocline. A slight increase in pH above the lake bottom in autumn indicates a reduction of iron oxihydroxides.

Iron and manganese-

Figure 5 shows the typical time-depth distribution of iron (Philippe 1989, Balistrieri *et al.* 1992). In sufficiently reducing environments, Fe(III) oxihydroxides are reduced and release Fe(II) to the pore water and into the water column. Concentration of manganese, another easily reducible element (Mn(IV) to Mn(II)), increases shortly before iron reaches 10 μM (fig. 6).

Nutrients and silica -

Orthophosphate concentrations are permanently below 0.5 μM except in autumn when they can reach 1.5 μM in the hypolimnion. Iron and phosphorus cycles in lakes are intimately related (O' Melia, 1985) and orthophosphate release during stratification could be limited by adsorption onto settling or authigenic oxihydroxide particles. Variation in nitrate concentrations with depth occurs sharply at the same period. Concentrations are lower than 0.5 μM, however, in October, a maximum of 2 μM occurs at a 22 m depth. The low surface values are caused by nitrate incorporation by the biota during organic matter (OM) production, whereas the low nitrate values in bottom waters can be a result of denitrification. The two distinctive features in the distribution of ammonium are the high (4 μM) concentrations throughout the water column in late November due to the beginning of the cold season overturn and the higher (up to 45 μM) concentrations that most likely develop as a result of organic matter diagenesis in the bottom waters and sediments (fig. 7).The larger concentrations of dissolved silica in the hypolimnion (fig. 8) from spring to early winter are most likely due to dissolution of Diatom skeletons.

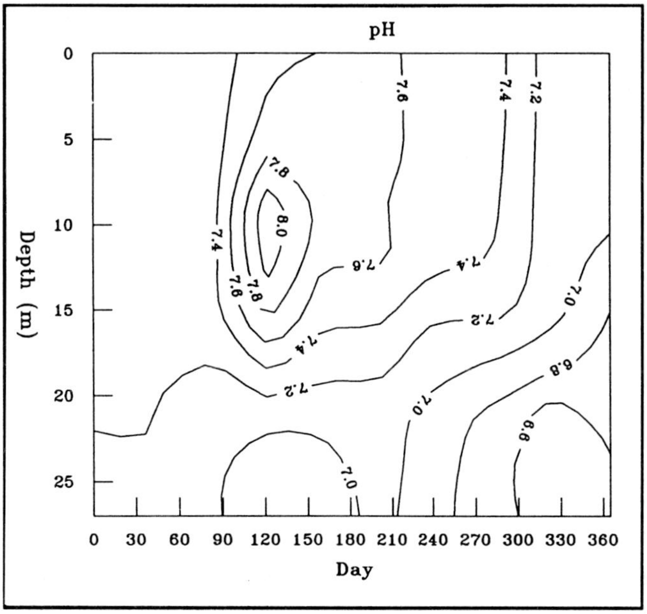

Figure 4 - pH isocontours in a time-depth graph

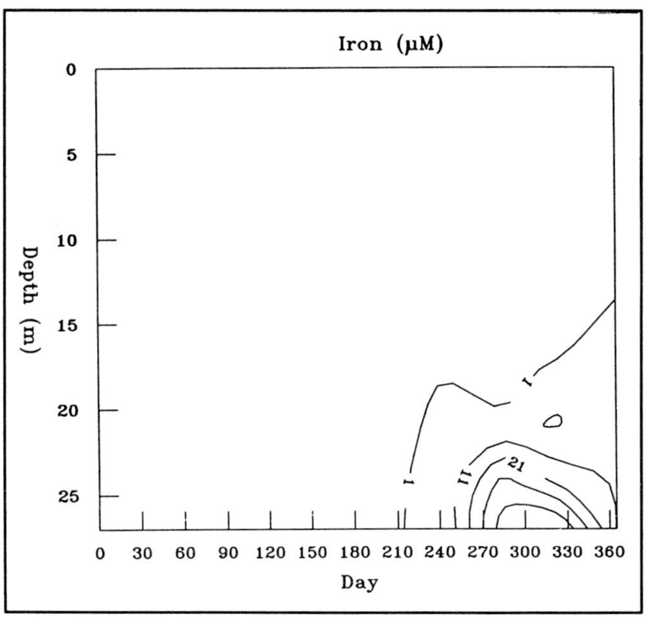

Figure 5 - Iron isocontours in a time-depth graph

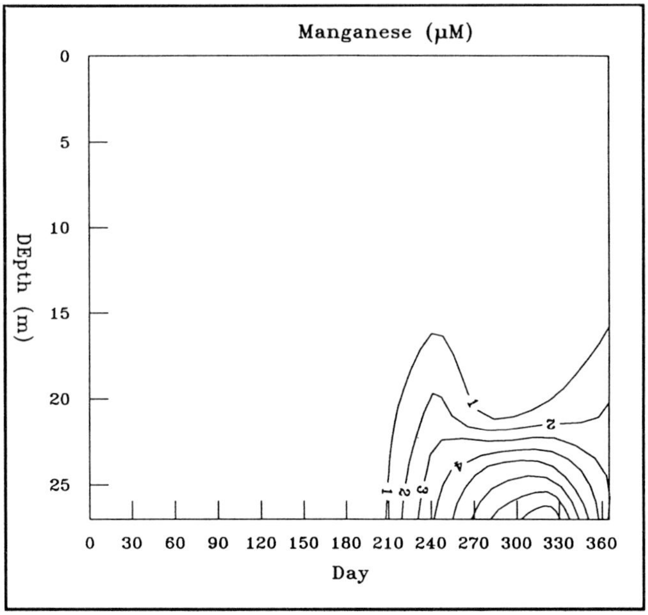

Figure 6 - manganese isocontours in a time-depth graph

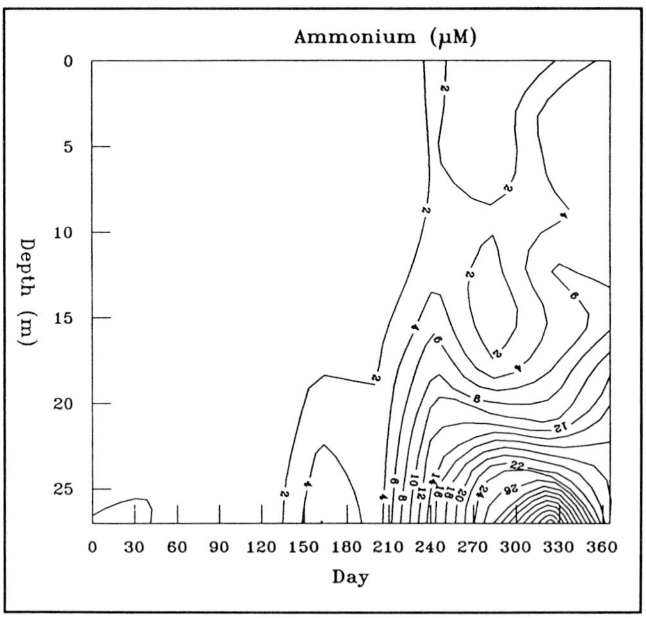

Figure 7 - Ammonium isocontours in a time-depth graph

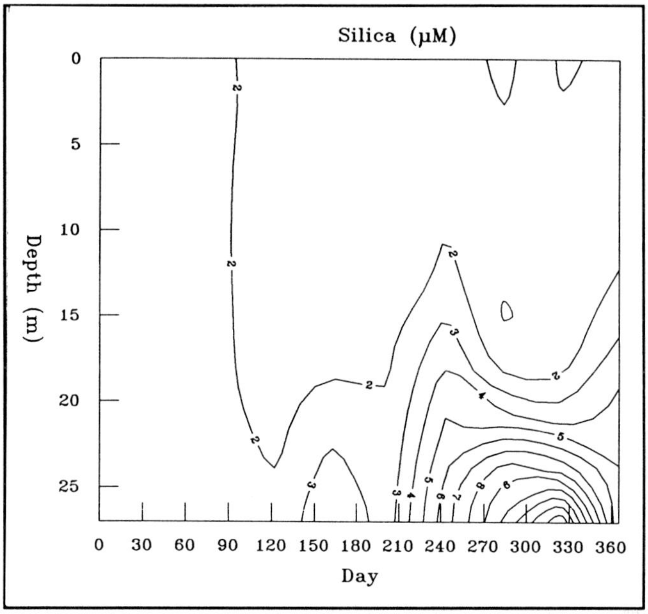

Figure 8 - Dissolved silica isocontours in a time-depth graph

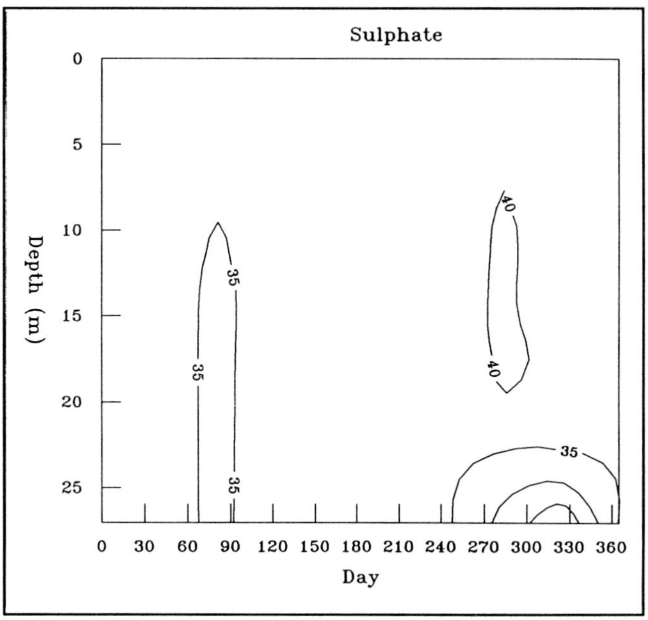

Figure 9 - Sulphate isocontours in a time-depth graph

Sulphate and organic carbon-

Reduction of sulphate in sulphide below 20 m happens when oxygen is below 100 μM (fig. 9). DOC is quasi-constant through the year with values around 170 ± 40 μM except May and June maxima (260-430 μM) observed in subsurface (2 m). POC ranges from 15 to 80 μM (fig. 10) and shows different patterns for increasing stratification.

Alkalinity, other major compounds and strontium-

Alkalinity concentrations range from 100 to 395 μM during the year. Highest concentrations occur near the bottom in November. Other dissolved compounds mean concentrations are summarised in Table 2 since their values do not show large variations during the year although very slight increases above the water-sediment interface for alkaline-earth could be coupled with redox conditions (Sholkovitz, 1985). Profiles are in agreement with data from Truze (1990) and did not show lateral variations.

Trace elements-

Semiquantitative analysis (analytical precision is ±20%) for the four following elements allows to distinguish between chemical release or scavenging in the water column during the year. Barium, cobalt, arsenic are released in the hypolimnion while vanadium is trapped in the reducing zone (fig. 11, 12, 13 and 14). From November 1993 sampling and quantitative analysis, we observe through the vertical profile : 1) non-reactive behaviour for boron, 2) a decrease then an increase for vanadium and aluminium, 3) an increase for barium, cobalt, arsenic, rubidium, molybdenum, caesium, cerium, and lead. Lithium, chromium, nickel, copper, zinc, cadmium, tungsten and uranium concentrations were below ICP-MS detection limits.

Springs and rainfall-

Ground waters and rainfall are potential reservoirs to feed the Lac du Bouchet. Mean concentrations concerning 8 sampled springs and mean concentrations for collected rainfall (isolated and cumulated samples) are presented in Table 2.

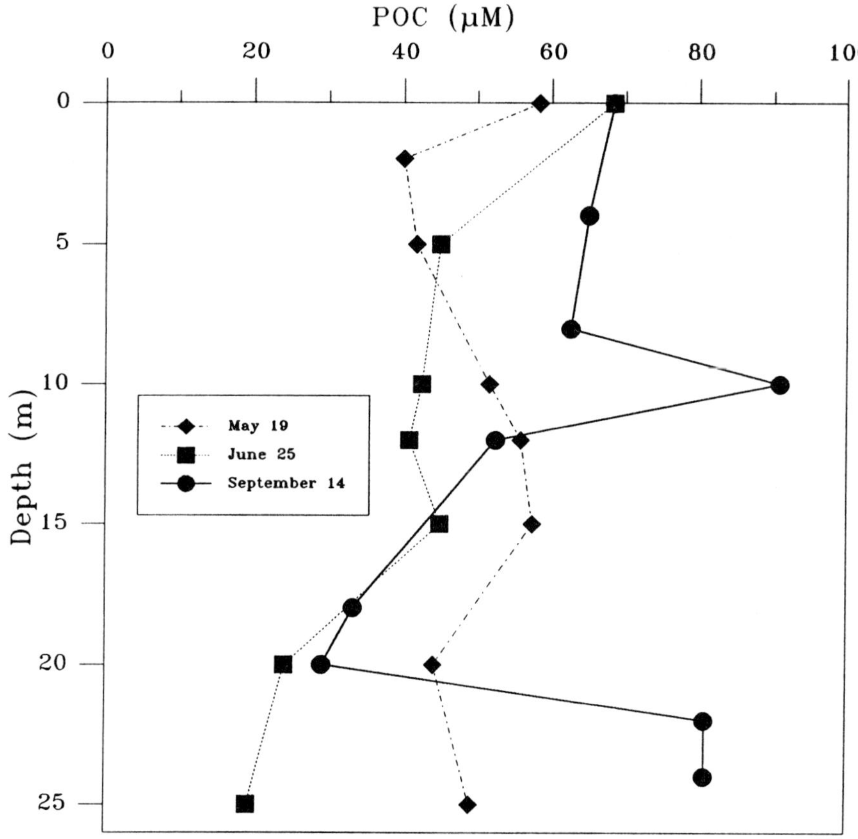

Fig. 10 – Particulate organic matter concentration vs. depth

Table 2: Concentration of the different components in springs, rainfall and the lake waters (Data are given in µM)

	Springs (n=8)		Rainfall (n=8)		Lake (n=144)	
Comp.	m_s	s	m_r	s	m_l	s
Chloride	107	72	16	12	37	3
Sulphate	66	33	21	11	37	3
Sodium	171	74	15	15	41	3
Potassium	51	83	6	4	11	1
Magnesium	262	30	10	8 (n=2)	61	4
Calcium	268	95	22	17	53	9
Silica	369	65	2.5	1.4	2.4	2.2
Strontium	2.0	1.0	0.07	0.07	0.23	0.02

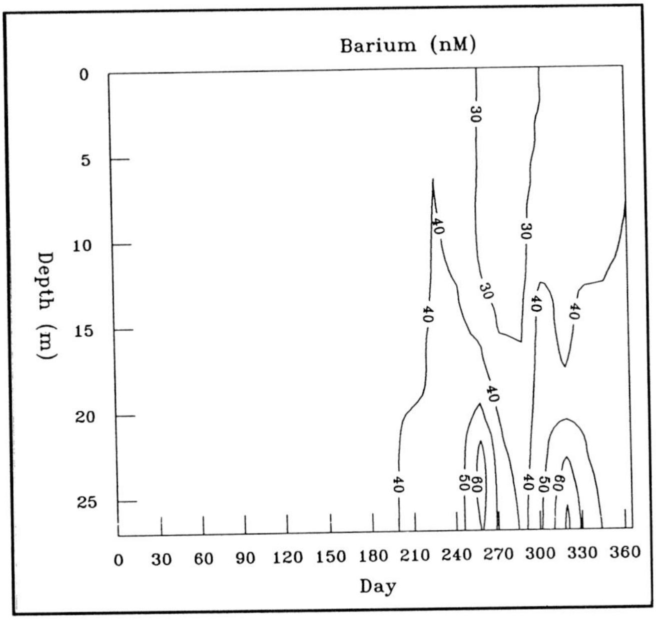

Figure 11 - Barium isocontours in a time-depth graph

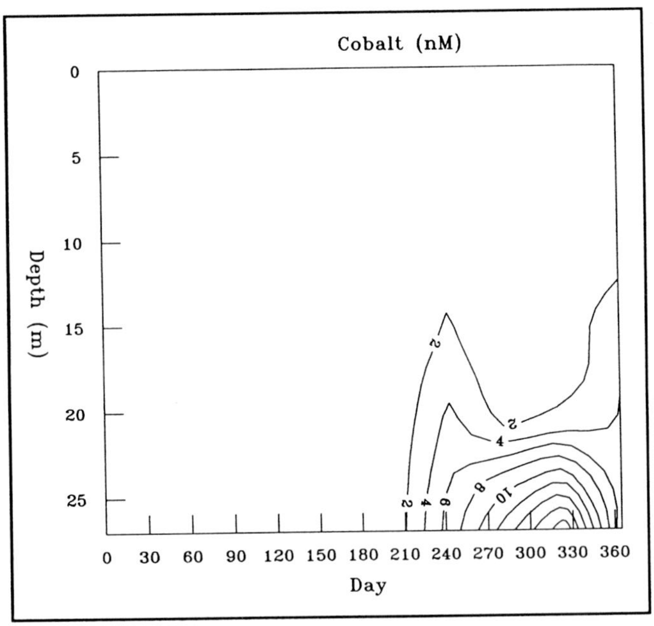

Figure 12 - Cobalt isocontours in a time-depth graph

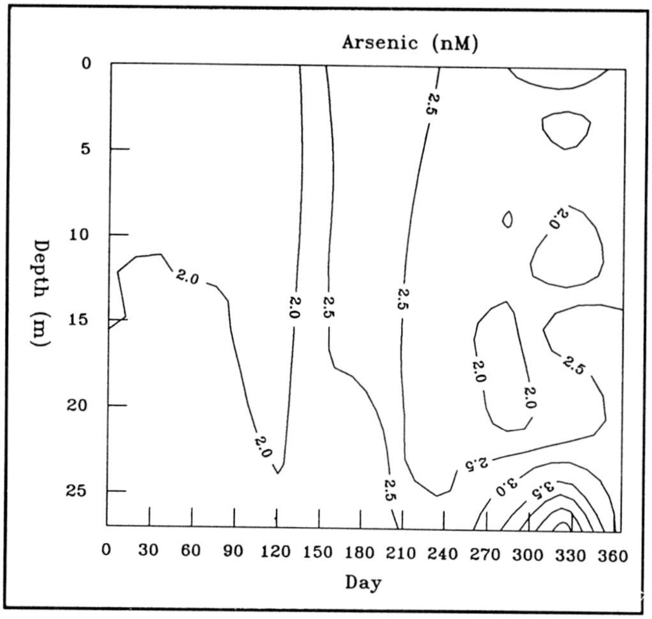

Figure 13 - Arsenic isocontours in a time-depth graph

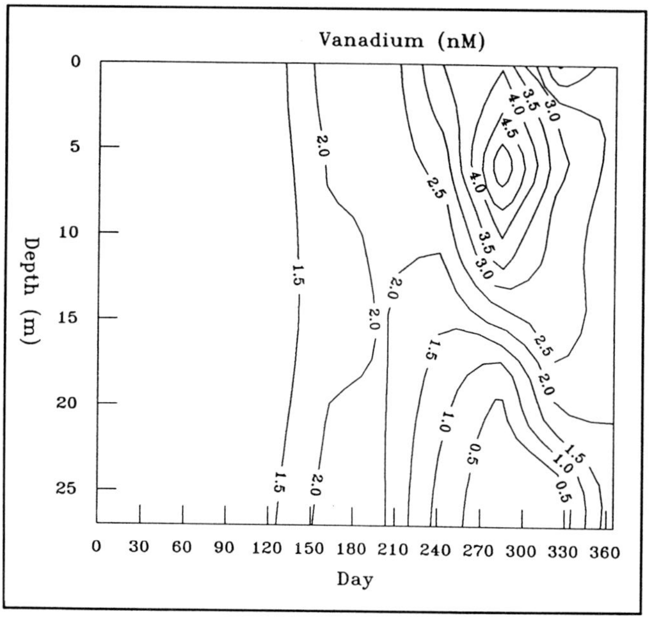

Figure 14 - Vanadium isocontours in a time-depth graph

Discussion-

Water balance-

With an annual tidal range of ± 30 cm, the lake level is described as unvarying from year to year (Casta 1991). Therefore most people who live near the Lac du Bouchet think it is fed by underlacustrine springs. From the first observation, we can consider that a hydrological steady state is established for a yearly well-mixed box. Accordingly, water inputs are equal to water outputs and allow us to write the following balance equation :

$$R + S = I + E$$

where R is rainwater inputs (rainfall and runoff), S is underlacustrine spring inputs, I is lake bottom infiltration outputs, E is lake surface evaporation outputs. This can be normalised to R owing to rainfall quantification from Cayres pluviometer records :

$$1 + S' = I' + E' \qquad S'=S/R \; ; \; I'=I/R \; ; \; E'=E/R$$

Furthermore, we can extend it to a chemical balance for conservative or semiconservative substances. These compounds should not be related significantly with the organic carbon cycle and concentrations in respective should be highly distinguishable. For dissolved compounds, the chemical balance can be written:

$$[a]_R + S'[a]_S = I'[a]_I + E'[a]_E$$

From 2 components a and b, it is possible to obtain a simple solution for S', I' and E'.

$[a]_E$ and $[b]_E$ negligible

$$S' = \frac{[a]_R[b]_I - [a]_I[b]_R}{[a]_I[b]_S - [a]_S[b]_I} \qquad I' = \frac{[a]_S[b]_R - [a]_R[b]_S}{[a]_S[b]_I - [a]_I[b]_S}$$

$$E' = 1 + S' - I'$$

where $[\ ]_R = m_r$, $[\ ]_S = m_S$ and $[\ ]_I = m_I$ (see Table 2).

Table 3: Calculations of S', I' and E' for various element (a, b) couples.

(a,b)	S'	I'	E'
(Cl, SO_4)	0.10	0.73	0.37
(Cl, Na)	0.04	0.55	0.49
(Cl, Ca)	0.03	0.51	0.52
(Cl, SiO_2)	0.00	0.42	0.58
(Cl, Sr)	0.02	0.49	0.55
(SO_4, Na)	0.07	0.67	0.40
(SO_4, Ca)	0.06	0.65	0.41
(SO_4,SiO_2)	0.00	0.55	0.45
(SO_4, Sr)	0.03	0.61	0.42
(Na, Ca)	-0.03	0.27	0.71
(Na, SiO_2)	0.00	0.37	0.63
(Na,Sr)	0.01	0.43	0.59
(Ca,SiO_2)	0.00	0.37	0.63
(Ca,Sr)	0.02	0.46	0.56
(SiO_2,Sr)	-0.01	0.27	0.72
mean	**0.02**	**0.49**	**0.54**
s	0.03	0.14	0.11

Table 3 presents calculations done for different element couples. If an underlacustrine feeding exists, its contribution is not important (2% of total inputs) but infiltration seems to be relatively high (48% of total inputs). Inputs are calculated from : 1) an experimental study in the Devès area (10 km from the Lac du Bouchet, Lecocq 1987) which gives an estimate for the annual efficient rain (Rainfall - Evapotranspiration) of 300 mm/y and 2) with the annual mean precipitation amounts derived from the last 30 years records (910 ± 90 mm/y). For the Bouchet forested drainage basin, the efficient rain value needs to be largely decreased to take account of direct evapotranspiration on leaves and roots uptake ; 100 (± 100) mm/y seems to be more reasonable. Finally, we find that lake composition is mainly rainwater. Evaporation is on the same scale of size (E/R corrected from runoff » 0.7) as the value found by Bouchet (1987) for the Lac d'Aydat (E/R » 1) since the Lac du Bouchet is 350 m more elevated. Fig. 15 shows an overview of the Lac du Bouchet hydrological functioning. Water residence time (q_w) is close to 14 years (± 4 years owing to the different sources of error).

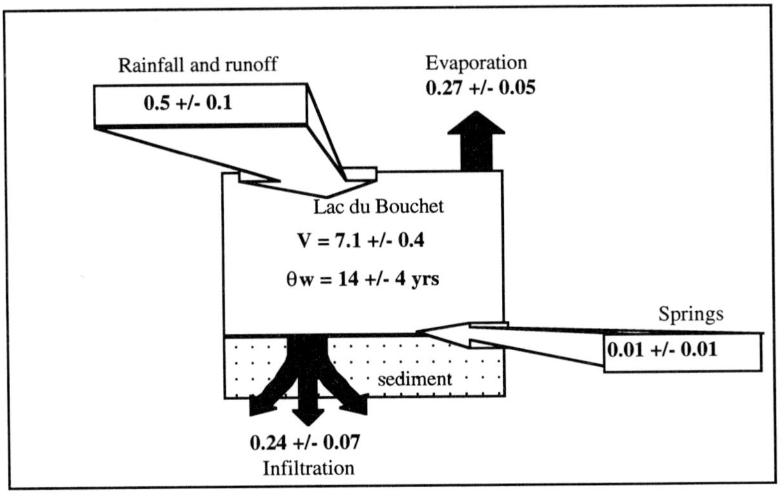

Figure 15: overview of the Lac du Bouchet hydrological functioning
(**V** is in 10^6 m^3 - Others are in 10^6 m^3/an)

Biogeochemical response-

Surface inflow is limited. Therefore, nutrients used during photosynthesis are mainly supplied, in surface lake water, by precipitation, atmospheric particles and hypolimnetic mixing (recycling during the stratification period). Despite this low nutrient feeding emphasised by the bottom seepage, concentration gradients in lakes, for dissolved compounds, are explained by OM production and its oxidation under bacterial mediation. Concentration profiles provide information regarding the advancement of chemical reactions in the water column. This gives rise to zones of a successively more reducing character with depth, as the available electron acceptors are consumed by bacteria in order of their thermodynamic advantage (Stumm and Morgan, 1981):

$$\text{"}CH_2O\text{"} + O_2 \longrightarrow CO_2 + H_2O \qquad \text{(a)}$$

$$\text{"}CH_2O\text{"} + 4/5\ NO_3^- + 4/5\ H^+ \longrightarrow CO_2 + 2/5\ N_2 + 7/5\ H_2O \qquad \text{(b)}$$

$$\text{"}CH_2O\text{"} + 2\ MnO_2 + 4\ H^+ \longrightarrow CO_2 + 2\ Mn^{2+} + 3\ H_2O \qquad \text{(c)}$$

$$\text{"}CH_2O\text{"} + 4\ Fe(OH)_3 + 8\ H^+ \longrightarrow CO_2 + 4\ Fe^{2+} + 11\ H_2O \qquad \text{(d)}$$

$$\text{"}CH_2O\text{"} + 1/2\ SO_4^{2-} + 1/2\ H^+ \longrightarrow CO_2 + 1/2\ HS^- + H_2O \qquad \text{(e)}$$

$$\text{"}CH_2O\text{"} + \text{"}CH_2O\text{"} \longrightarrow CO_2 + CH_4 \qquad \text{(f)}$$

"CH_2O" is the simplified OM formula.

Figure 16 presents an inorganic carbon path during stratification as a function of oxygen level. Three features on this graph can be described to understand the main biogeochemical processes:

1) samples from epilimnion can be represented by a straight line with a slope close to zero meaning these waters are in equilibrium with atmospheric O_2 and CO_2 at constant alkalinity,

2) linear variation in the hypolimnion with a ≈ -1 slope is due to the (a) reaction [if photosynthesis was significant, epilimnion data should give line intermediate between the lines 1) and 2)],

3) the small vertical variation of [ΣCO_2], when [O_2] is close to 0, is due to the (b), (c), (d), (e) and (f) reactions.

In the Lac du Bouchet, the latter reactions account for less than 10 % (maximum November value) of the inorganic carbon liberated just over the bottom (fig. 17, 18, 19 and 20). This value must be considered as a maximum since diffusion from interstitial water is not taken into account. Diffusion is also a source of reduced compounds as Fe^{2+}. Such species can precipitate as oxihydroxides in the suboxic hypolimnion and scavenge different elements from the water column (including humic substances, Tipping and Woof, 1983). This leads to an explanation for the peak of POC in the September profile (fig. 10) below 20 m.

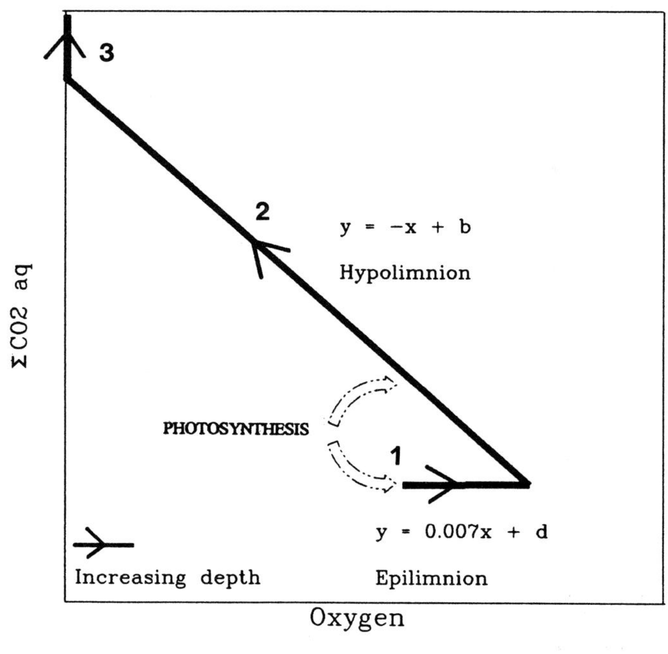

Fig. 16 – Inorganic dissolved carbon path in the lake – Theoretical diagram

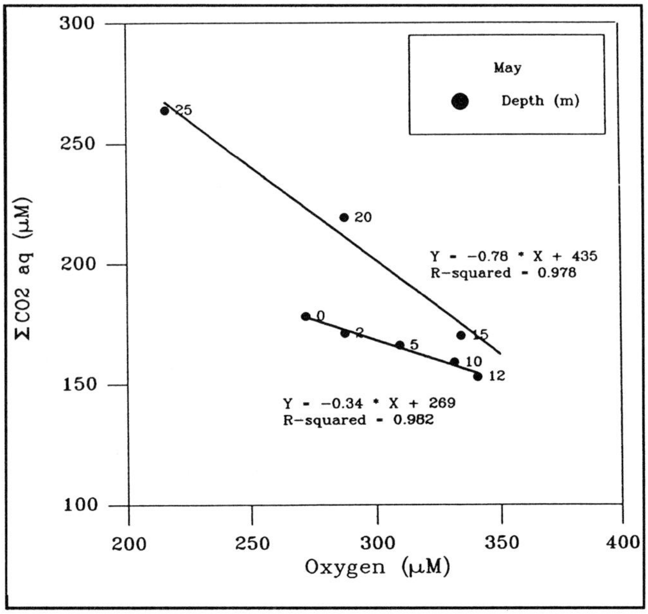

Figure 17 - Inorganic dissolved carbon path in the lake - May data

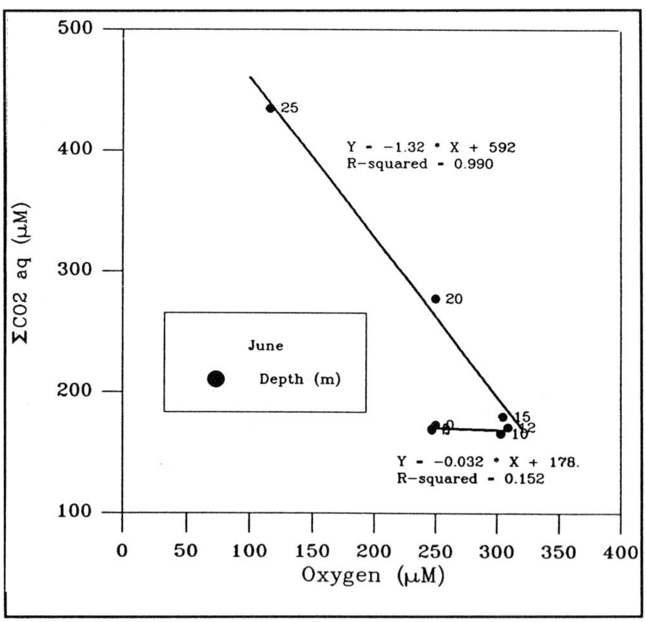

Figure 18 - Inorganic dissolved carbon path in the lake - June data

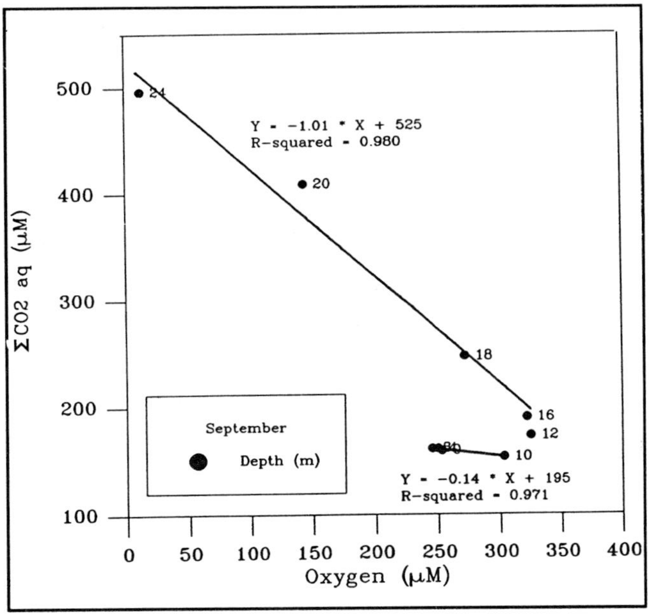

Figure 19 - Inorganic dissolved carbon path in the lake - September data

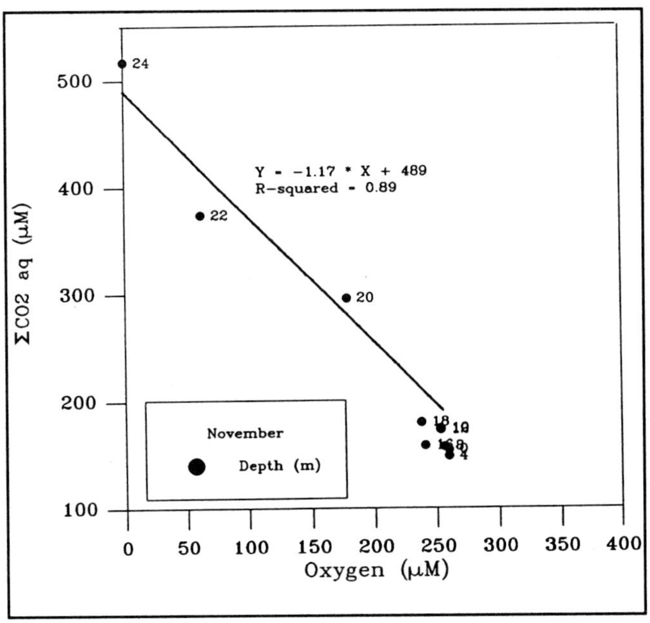

Figure 20 - Inorganic dissolved carbon path in the lake - November data

Trace element distribution-

Later in autumn, when the oxidation boundary layer slightly rises up in the water column, reduced species are more stable under anoxic conditions. They can diffuse with Fe^{2+} and enrich the bottom water (the source of mobilisation reactions is located in the sediment). It is the case of various chemical elements : Co, As, Mo, V, Ce, Pb, and Al (fig. 21). For Mn and Ba (fig. 22), source reactions are linked together and could be described in the water column (Sugiyama, 1992) as adsorbed Ba liberation below the redox interface during Mn oxihydroxides reduction (reaction c). Al variations can without doubt be related to a pH change in the lake (fig. 23) which could involve dissolution of Al-containing minerals (from atmospheric inputs) such as clays or hydroxides in particulate and colloidal forms (Marin Galvin, 1991) since Al solubility increases for pH above 7. Significant higher V concentration in the upper part of the lake (Autumn) may also result from a pH sensitive solid phase and its scavenging below the redox interface could be due to a lower solubility in anoxic media (Prange and Kremling, 1985). It seems to be also the case for Sb and Cu (fig. 24). This means that in anoxic conditions, elements such as V or Mo (Viollier *et al.*, in preparation) can be both incorporated in solid phase and stabilised in solution, probably related to the Fe and OM cycles.

Conclusion-

An explanation for the non evolved geochemical signature of the Lac du Bouchet (oligotrophic type) may be deduced from the chemical balance which suggests the output seepage of lake waters throughout the lake bottom. Photosynthesis and OM degradation are significant enough to create a fairly oxygen depleted hypolimnion. OM biodegradation in the water column is carried out mainly by O_2, but even if oxidation with O_2 is more efficient, the quantitatively important biogeochemical processes occur in the anoxic sediment (Jézéquel *et al.*, this volume). Interstitial waters are the location both of Fe oxihydroxides reduction and various trace element release which can diffuse in the November anoxic bottom water.

Acknowledgements- Fieldwork was carried out with the help of the Laboratoire de Géologie du Quaternaire (université d'Aix-Marseille II). Emile Thiebault provided help with collecting rainfall samples. Meteorological data come from Météo France Haute-Loire (station de Loudes).

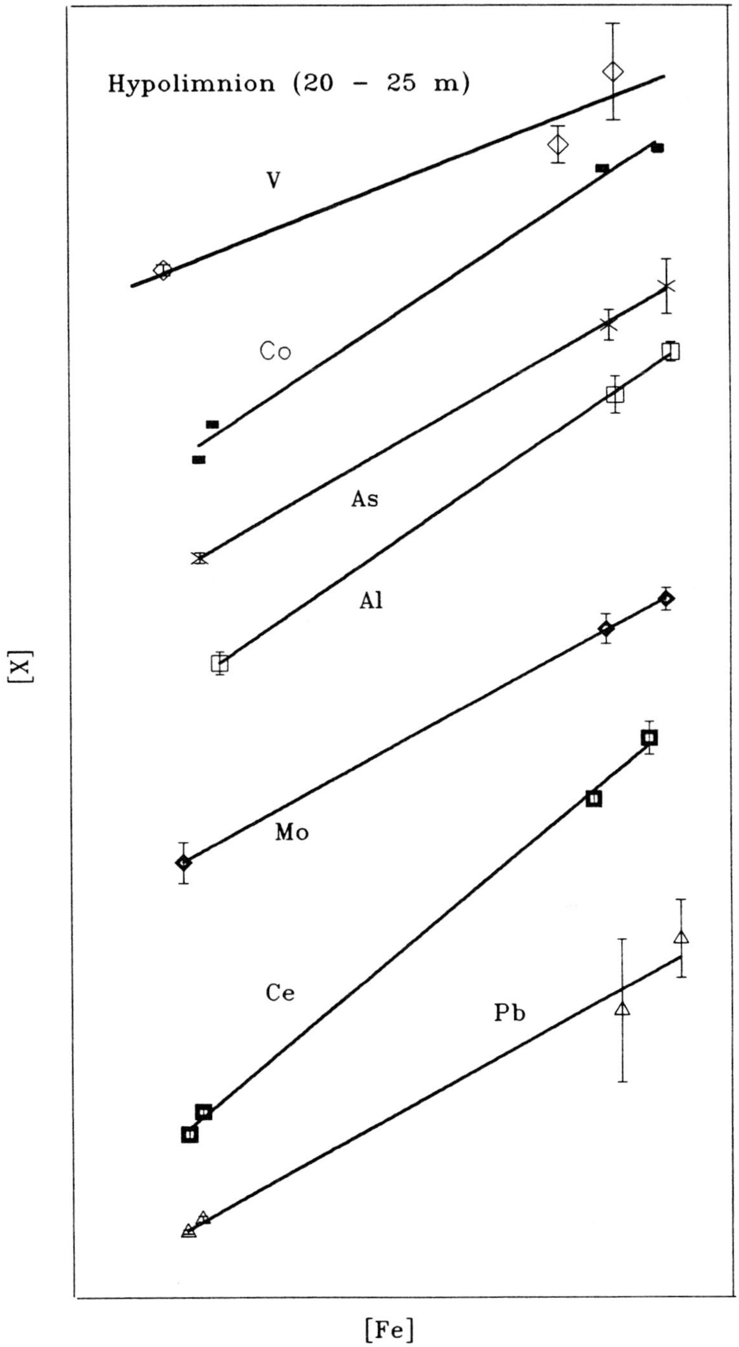

Fig. 21 – Trace elements concentration ([x] arbitrary unit) vs. iron concentration (a.u.)

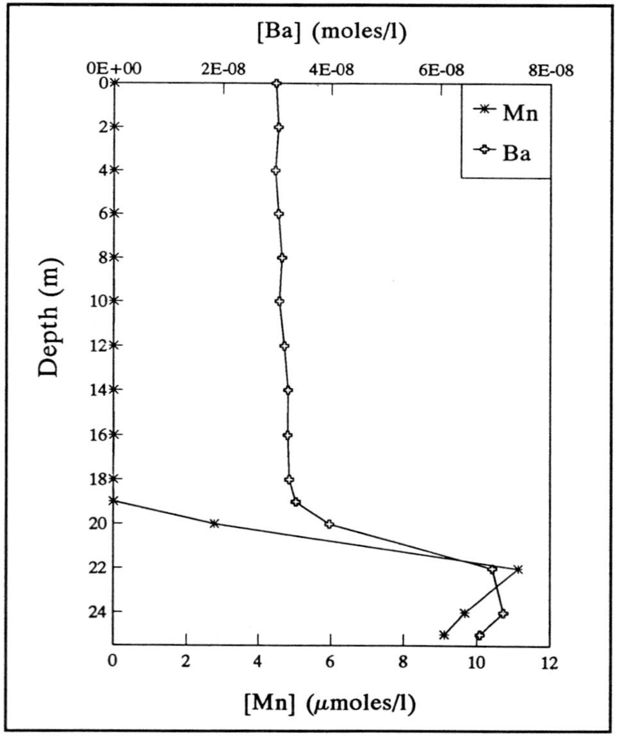

Figure 22 - Barium and manganese profiles (November 1993)

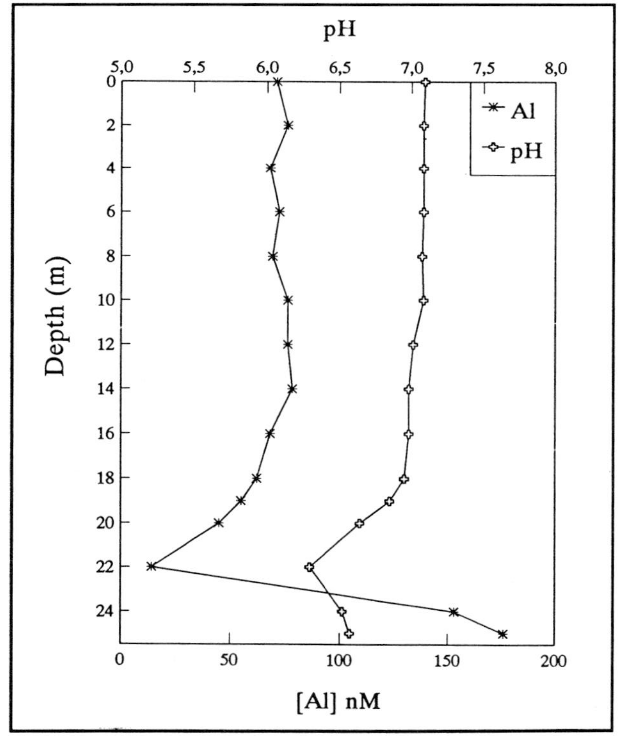

Figure 23 - pH and Aluminium profiles (November 1993)

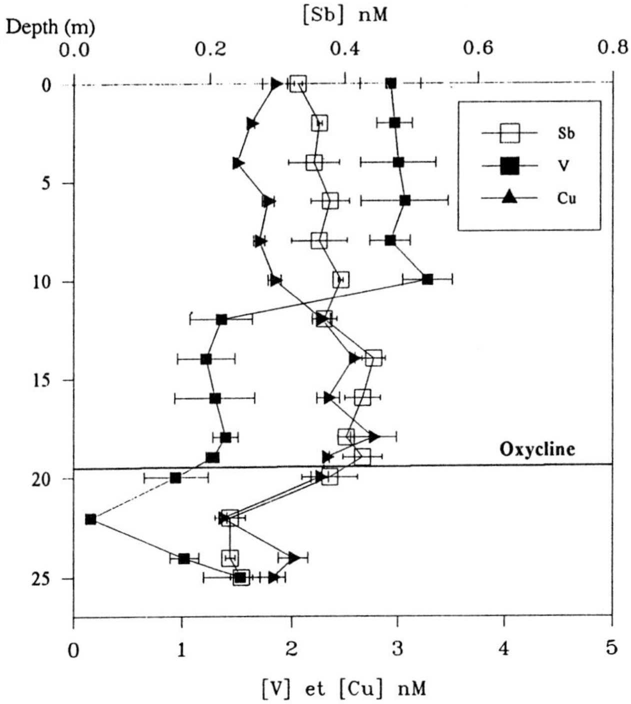

Fig. 24 – Antimony, vanadium and copper profiles (November 1993)

References-

Balistrieri L.S., Murray J.W. and Paul B. (1992) The cycling of iron and manganese in the water column of Lake Sammamish, Washington *Limnol. Oceanogr.*, **37**(3), 510-528.

Bouchet C. (1987) Hydrogéologie du milieu volcanique. le bassin de la Veyre - Thèse de 3ème cycle - Université d'Avignon

Casta L. (1991) Les structures thermométriques et pHmétriques : mise en oeuvre, implications hydrologiques, climatiques et sédimentologiques *In* "Le Lac du Bouchet: environnement naturel et étude des sédiments du dernier cycle climatique". Documents du C.E.R.L.A.T. n° 2, E. BONIFAY (ed.), pp.97-111.

Eckartz-Nolden G. and Nolden M. (1991) Lac du Bouchet (France, Massif Central) : results of two investigations : chemistry, phytoplankton and zooplankton- *In* "Le Lac du Bouchet: environnement naturel et étude des sédiments du dernier cycle climatique". Documents du C.E.R.L.A.T. n° 2, E. BONIFAY (ed.), pp. 79-88.

Gran G. (1952) Determination of the equivalence point in potentiometric titration part. II - International Congress on Analytical Chemistry, **77**, 661-671.

Lecocq A. (1987) Hydrogéologie en milieu volcanique - Etude de la partie nord du plateau basaltique du Devès. Thèse de 3ème cycle, Université Blaise Pascal (Clermont-Ferrand II), 221 p.

Marin Galvin R. (1991) Study of evolution of aluminium in reservoirs and lakes. *Wat. res.*, **25**(12), 1465-1470

Mergoil J. (1987) Aperçu géologique du Velay. *Doc. du CERLAT*, mém. n°1, 17-22

Milligan C.W. and Lindstrom F. (1971) Colorimetric determination of calcium using reagents of the glyoxal bis (2- hydroyanil) class. *Anal. Chem.*, **44**, 1822-1826.

O' Melia C.R. (1985) The influence of coagulation and sedimentation on the fate particles, associated pollutants and nutrients in lakes. *In* Chemical processes in lakes, STUMM W.(ed.), Wiley interscience, 435 p.

Philippe L. (1989) Bilan géochimique du fer et du phosphore dans un écosystème lacustre eutrophe : le lac d'Aydat (Puy de Dôme). Thèse de 3ème cycle, Université Paris 7. 186 p.

Prange A. and Kremling K. (1985) Distribution of dissolved molybdenum, uranium and vanadium in Baltic sea waters. *Mar. chem.*, **16**, 259-274

Sholkovitz E. (1985) Redox related geochemistry in lakes : alkali metals, alkaline earth elements and ^{137}Cs. *In* Chemical processes in lakes, STUMM W. (ed.), Wiley interscience, 435 p.

Stumm W. and Morgan J.J. (1981) Aquatic chemistry, 2nd edition. Wiley interscience, 780p.

Sugiyama M., Toshitaka H., Sorin K. and Masakazu M. (1992) A geochemical study on the specific distribution of barium in Lake Biwa, Japan. *Geochim. Cosmochim. Acta*, **56**(2), 597-605.

Teulade A., Mergoil J. and Boivin P. (1991) Etudes géologique et volcanologique des environs du Lac du Bouchet - *In* "Le Lac du Bouchet: environnement naturel et étude des sédiments du dernier cycle climatique". Documents du C.E.R.L.A.T. n° 2, E. BONIFAY (ed.), pp.63-78

Tipping E. and Woof C. (1983) Elevated concentrations of humic substances in a seasonally anoxic hypolimnion: evidence for co-accumulation with iron. *Arch. Hydrobiol.,* **98**(2), 137-145.

Truesdale V.W. and Smith C.J. (1975) The formation of molybdosilicic acids from mixed solutions of molybdate and silicate. *Analyst*, **100**, 203-212

Truze E. (1990) Etude sédimentologique et géochimique des dépôts de maar du Bouchet (Massif Central, France) - Evolution d'un système lacustre au cours du dernier cycle climatique (0-120 000 ans). Thèse de 3ème cycle, Université d'Aix-Marseille II. 242 p.

U.S.E.P.A. (1984) Method for chemical analyses of water and wastewater - EPA 600/4-79-020-Nitrogen, ammonia-method 350.1 (colorimetrics, automated phenate) - STORET n°00610, dissolved 00608.

Whitledge T.E., Malloy S.C., Patton C.J. and Winick C.D. (1981) Automated nutrient analyses in seawater, technical report, Brook-haven National Laboratory, Upton, NY. 48p.

Geochemical study of the Lac du Bouchet (Hte-Loire, France) Part II : water - sediments - organic matter interactions during the last 2500 years

Didier Jézéquel[1], Patrick Albéric[2], Alain Desprairies[3], Marc Evrard[1], Dominique Lavergne[1], Gil Michard[1], Andrew J. Patience[2], Monique Pepe[1], Gérard Sarazin[1], Nicolas-Pierre Tribovillard[3] and Eric Viollier[1]

1) Laboratoire de Géochimie des Eaux, Université Paris VII, case postale 7052, F-75251 Paris cedex 5
2) Université d'Orléans, URA 724 du CNRS, Dépt. des Sciences de la Terre, F-45067 Orléans cedex
3) Université Paris Sud, URA 723 du CNRS, bâtiment 504, F-91405 Orsay cedex

Key words- *Maar lake, Lac du Bouchet, lacustrine sediments, interstitial water, early diagenesis.*

Abstract- The early diagenesis of the superficial sediment of the Lac du Bouchet has been studied by analysis of interstitial water and solid phase. ^{14}C dating of the sediment gives 2500 ± 350 years at a depth of 80 cm below the sediment-water Interface (SWI). Solid phase is made of 19 to 49% clays (mainly kaolinite), detrital minerals (quartz, feldspars, ferromagnesian), Diatoms frustules and organic matter (ligno-cellulosic debris, pollens, amorphous grey matter). The main part of organic matter (OM) mineralisation takes place in the first 10 to 20 centimetres under the SWI, but continues at a deeper level 3 to 4 times more slowly. Methanogenesis reaction is the dominant degradation process (90 to 93% in the upper section of the mud, up to 97% below), followed by sulphate reduction (4% in spring and 7% in autumn, only in the top section) and iron reduction (3 to 2%). A formula is proposed for OM undergoing oxidation : $(CH_2O)_{106}(NH_3)_{13.6}(H_3PO_4)_{0.84}$.

Introduction-

The Lac du Bouchet (Massif Central, France) is a well studied lacustrine environment, particularly since the EUROMAARS EEC program started (Bonifay and Truze, 1987 and 1991). This maar lake presents a remarkable continuous sedimentation which has gone on for about 750,000 years. During this period, 60 metres of sediments have been accumulated (Teulade *et al.*, 1991). Palaeoclimatic investigations were performed up to the last 350,000 years, but only a few data have been published about the recent sediment layer (Bertrand *et al.*, 1992, Lallier-Vergès *et al.*, 1993, Patience *et al.*, this volume). This paper presents a geochemical investigation of the very superficial sediment (1 m) from an early diagenesis point of view.

The Lac du Bouchet is located on a volcanic plateau (Devès) at an altitude of 1205 metres above the sea level. The lake has neither visible inlet nor visible outlet : rain water seems to be the main water supply (Truze, 1990 ; Viollier *et al.*, this volume). The low sedimentation rate (0,1 to 0,3 mm/y) results from the poor washing of the wooded catchment area and from oligotrophic statute of the lake water.

Sampling-

The deepest point of the lake is 28 m below the surface so samples of sediment and samples of pore water were done by scuba diving. We preferred this way rather than sampling from a boat (with a Mackereth corer for example) because of the high fragility of the top sediment layer.

Eight sediment cores were extracted in the central part of the basin (figure 1) with 20 to 80 cm PVC sharpened-end tubes. Nitrogen was insuflated into the top of the cores to prevent oxidation and an isotherm container was used for transport.

Pore water was sampled by a dialysis technique (Carignan, 1984) which is a suitable sampling method for such a soft material. Dissolved compounds diffuse from sediment water through a porous membrane (Pall Biodyne Nylon 6-6, 0,2 µm porosity) to 20 ml compartments initially filled with desionised water. Four weeks after their implantation in sediment, dialysers (or peepers) were retrieved and water was quickly sampled, and protected from air and/or acidified (sediment was totally anoxic).

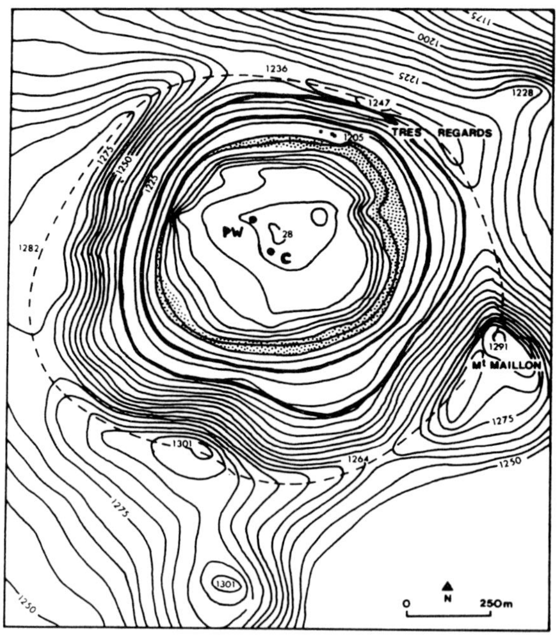

FIG. : 1 - Topographic and bathymetric map of the Lac du Bouchet, with localization of pore water (PW) and core (C) samplings.

Analytical methods-

Solid phase-

Granulometry, X-Ray spectroscopy, porosity (weight difference between a moist core slice and the same dry sample), ^{14}C dating.

Solid phase (exchangeable part)-

Sequential chemical attacks of sediment release compounds which are increasingly strongly bound to minerals or organic matter (elements into clayey layers or adsorbed on clays surface, complexed compounds, colloidal or poorly crystallised phase). Processing was:
1) NH_4NO_3 (1 mol/l) attack for an hour: exchangeable cations are released from clayey layers.
2) EDTA (0.05 mol/l) attack : release of adsorbed cations on clays and weakly complexed cations with organic matter.
3) NH_2OH-HCl (0.1 mol/l) and HNO_3 (0.01 mol/l) : release of more strongly bound cations.

Pore water composition-

Several non conservative compounds were analysed a few hours after sampling: pH (micro electrode, samples were protected from air contact), alkalinity by colorimetry (Podda and Michard, 1994), phosphate and iron (colorimetry with a Merck set), ammonia by colorimetry (Lange set). Other elements were analysed at the Laboratory : major anions by ionic chromatography (Dionex column, Shimadzu CDD-6A conductimeter), major cations by the same technique or by atomic absorption flame spectrometry (AAS, GBC 902), manganese, aluminium by atomic absorption spectrometry with a graphite furnace (GF-AAS, Hitachi 180-70), silica, ammonia, nitrate and phosphate by microcolorimetry (Autoanalyser Alpkem).

Dissolved organic carbon (DOC) and methane were determined with a TCM 140 Carlo Erba. Hydrophobic DOC (humic acids) were separated from DOC with a XAD 8 column at pH = 1.5 (Thurman and Malcolm, 1981).

Results and discussion-

Dating of sediment -

Superficial sediment (1 m to 1.5 m thick in the centre of the lake) was dated from present time to the beginning of the Holocene (Truze, 1990; Lallier-Vergès *et al.*, 1993). Our results of the dating we made of the core (LDB S3) sampled in October 1992 are 2500 ± 350 years at a depth of 50 cm (fig. 2). This means that the average sedimentation rate is 0.2 mm a year.

Porosity-

Surface porosity reaches 93%. It decreases rapidly down to 88% in the first 4 cm, then slower from 4 to 28 cm where the porosity is 78%. Below, porosity is variable and increases again at 50 cm level (fig. 3). These different silts could result from a slumping. This porosity is relatively low compared to other lacustrine sediments (for example, in Lac d'Aydat, surface porosity is 98% and 92% at a depth of 40 cm ; Philippe, 1989).

Solid phase composition-

Sediment is an organic-rich clayey sediment. Mineral fraction was qualitatively homogeneous in the 80 cm sampled (Patience *et al.*, 1994). It was essentially represented by detrital minerals:
- clays (19 to 49% vol. of solid material) :

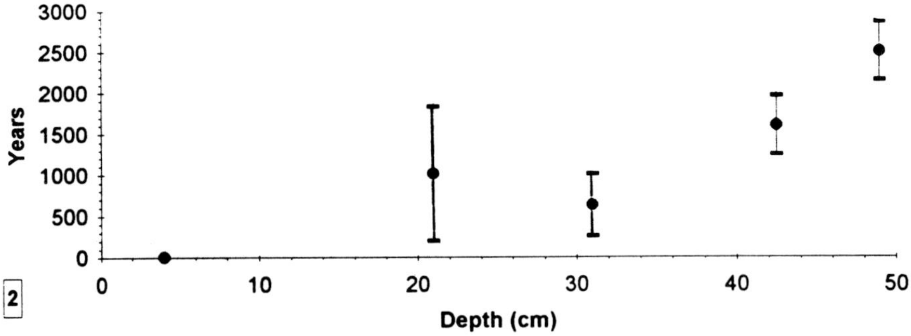

FIG. : 2 - 14C dating of sediment (core sampled in October 1992)

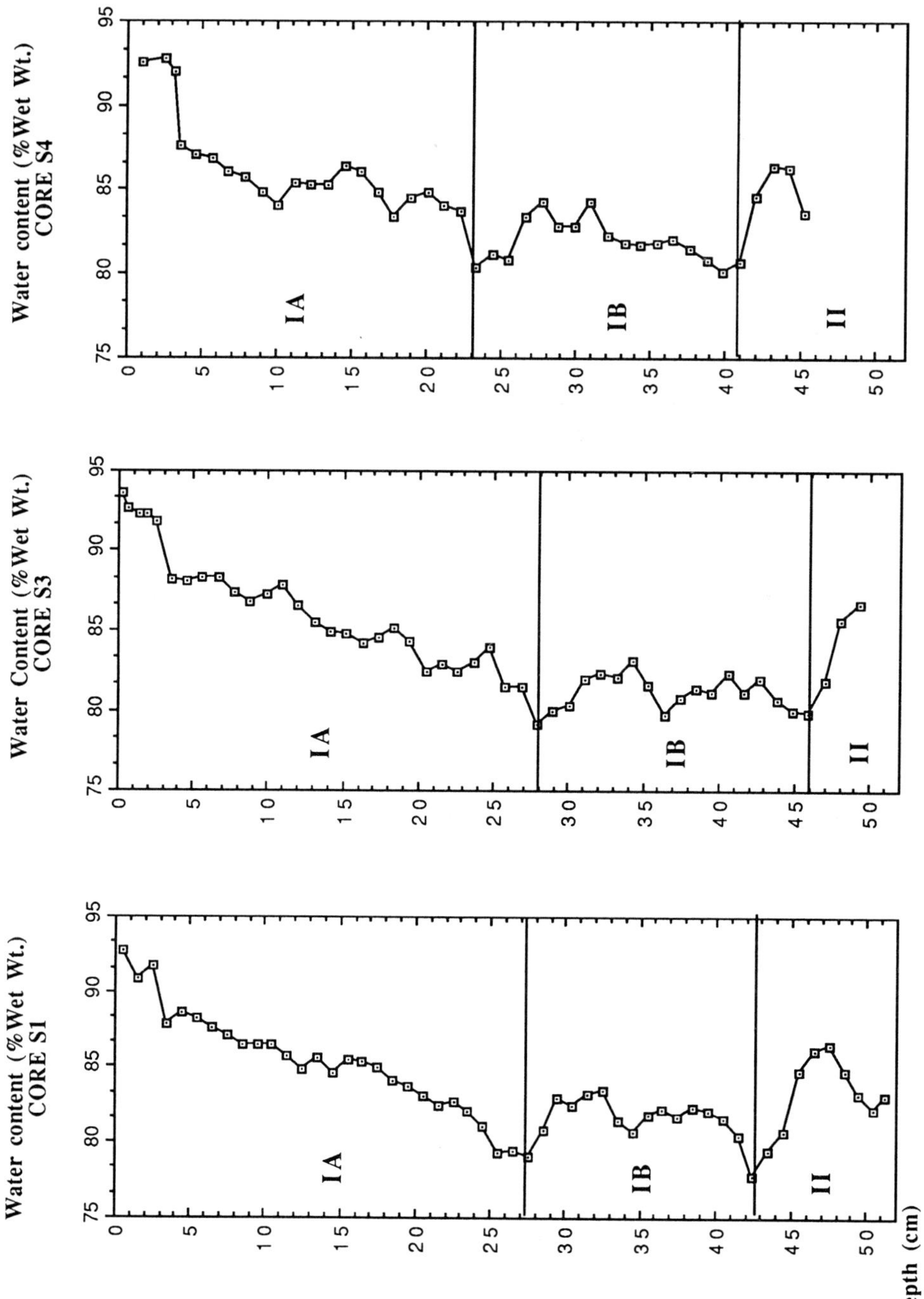

Figure 3: Porosity sections determined from 3 cores (October 1992)

kaolinite (the most important component of the clays, well crystallised, 2 to 6 µm) ; halloysite (poorly crystallised, 1 to 2 µm) ; chlorite (poorly crystallised, 1 to 6 µm); illite and smectite (plus gibbsite, 2 to 6 µm)

- quartz, feldspar, ferromagnesian (silt part from 6 to 40 µm and coarse part from 300 to 400 µm)
- bio-opal (Diatoms tests, 40 to 200 µm)
- pyrite FeS_2

Sediment was divided into four fractions :

- 1 : 1 - 6 µm : clays
- 2 : 15 - 40 µm : silt
- 3 : 40 to 200 µm : sand and organic fragments (Diatoms)
- 4 : > 200 µm : ligno-cellulosic debris and Diatom fragments.

The sediment was subdivided into three facies : IA (0 to 25 cm) and IB (25 to 40 cm), mostly made of clays and silt, and facies II, coarser. Clay/silt ratio was practically constant in the core, so the detrital supplies were constant also during this period.

Organic matter mineralisation-

The relative importance of organic matter (OM) mineralisation reactions can be estimated from the concentration gradients of produced or consumed species (concentration gradient for i will be noted Δi). In a first approximation, Redfield formula $(CH_2O)_{106}(NH_3)_{16}(H_3PO_4)$ was chosen for organic matter composition (C/N and N/P ratios are adjusted in part 5). In the anoxic sediment of the lake, the reaction sequence is :

(1) sulphate reduction and FeS precipitation :

$$OM + 53\ SO_4^{2-} + 53\ Fe^{2+} \longrightarrow 106\ CO_2 + 16\ NH_3 + H_3PO_4 + 53\ FeS + 106\ H_2O$$

$\Delta\Sigma CO_2 = 106/53\ \Delta SO_4^{2-}$; $\Delta alk = 15/53\ \Delta SO_4^{2-}$; and $\Delta NH_3 = 16/53\ \Delta SO_4^{2-}$

(2) manganese dioxide reduction :

$$OM + 212\ MnO_2 + 424\ H^+ \longrightarrow 106\ CO_2 + 16\ NH_3 + H_3PO_4 + 318\ H_2O + 212\ Mn^{2+}$$

$\Delta\Sigma CO_2 = 106/212\ \Delta Mn^{2+}$; $\Delta alk = 439/212\ \Delta Mn^{2+}$; and $\Delta NH_3 = 16/212\ \Delta Mn^{2+}$

(3) iron oxide reduction :

$$OM + 424\ Fe(OH)_3 + 848\ H^+ \longrightarrow 106\ CO_2 + 16\ NH_3 + H_3PO_4 + 1166\ H_2O + 424\ Fe^{2+}$$

$\Delta\Sigma CO_2 = 106/424\ \Delta Fe^{2+}$; $\Delta alk = 863/424\ \Delta Fe^{2+}$; and $\Delta NH_3 = 16/424\ \Delta Fe^{2+}$

(4) methanogenesis :

$$OM \longrightarrow 53\ CO_2 + 53\ CH_4 + 16\ NH_3 + H_3PO_4$$

$$\Delta\Sigma CO_2 = 53\ \Delta OM;\ \Delta alk = 15\ \Delta OM\ ;\ \text{and}\ \Delta NH_3 = 16\ \Delta OM$$

Nitrate was not detected in the water, so denitrification and nitrate reduction are not taken into account here.

a) Organic dissolved fraction.

Total DOC fluctuates from 2-3 mg/l at the sediment-water interface (SWI) to 5-6 mg/l near 1 metre depth, with a strong maximum near 10 cm under SWI in October 1992 (fig. 4a). This maximum may correspond to a faster degradation of organic matter or/and a locally decreased water-solid phase ratio. These values were lower in May 1993 (4.5 mg/l in the bottom), without any maximum near the SWI.
Methane was detected in the very first centimetres under the sediment-water limit (fig. 4b). Nevertheless, dissolved CH_4 and CO_2 were not measured out with a sufficient accuracy because of fast losses in atmosphere during sampling.
Hydrophilic DOC (not retained on XAD resins) in pore water was between 0.5 and 1.5 mg/l. No significant evolution was detectable in the section. Hydrophobic to hydrophilic DOC ratio was fluctuating, from 1 to 3 (fig. 4c)

b) Mineral dissolved fraction

Only conservative elements in the sampling conditions must be used in the model. Concentration gradients of main oxidiser agents allow calculation of the respective quantities of CO_2, alkalinity, NH_3 and H_3PO_4 released in pore water. The difference between the sum of all these quantities and the measured values are ascribed to the last process: methanogenesis.

Concentration sections in the sediment $Ci = f(z)$ are listed for major compounds in figures 5 to 15 (fig. 5: pH ; fig.6a : alkalinity and 6b : ΣCO_2, calculated with pH and alkalinity values;
fig. 7: Fe^{2+} ; fig. 8: Mn^{2+} ; fig. 9: NH_4^+ ; fig. 10: Na^+, K^+, Ca^{2+} and Mg^{2+} ; fig. 11: SO_4^{2-} ; fig. 12: Cl^- ; fig. 13: PO_4 ; fig.14: SiO_2 ; fig. 15: Al) and in table I (Oct. 1992) and II (May 1993).

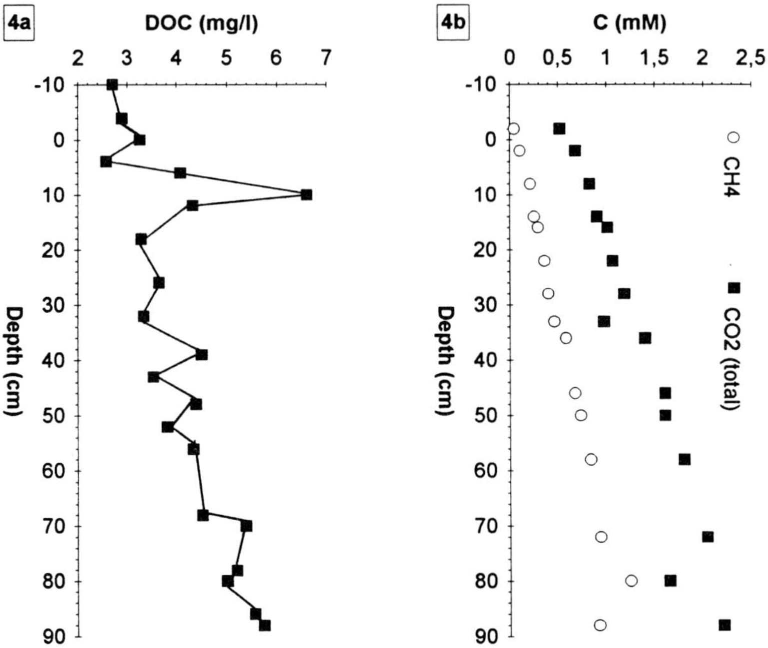

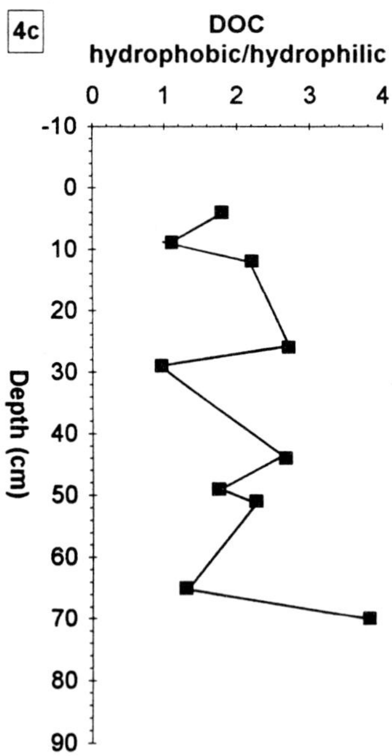

Fig. 4 : a - Dissolved organic carbon (DOC) section in October 1992. b - Methane and total mineral CO_2 (October 1992). c - DOC hydrophobic/hydrophilic ratio (May 1993).

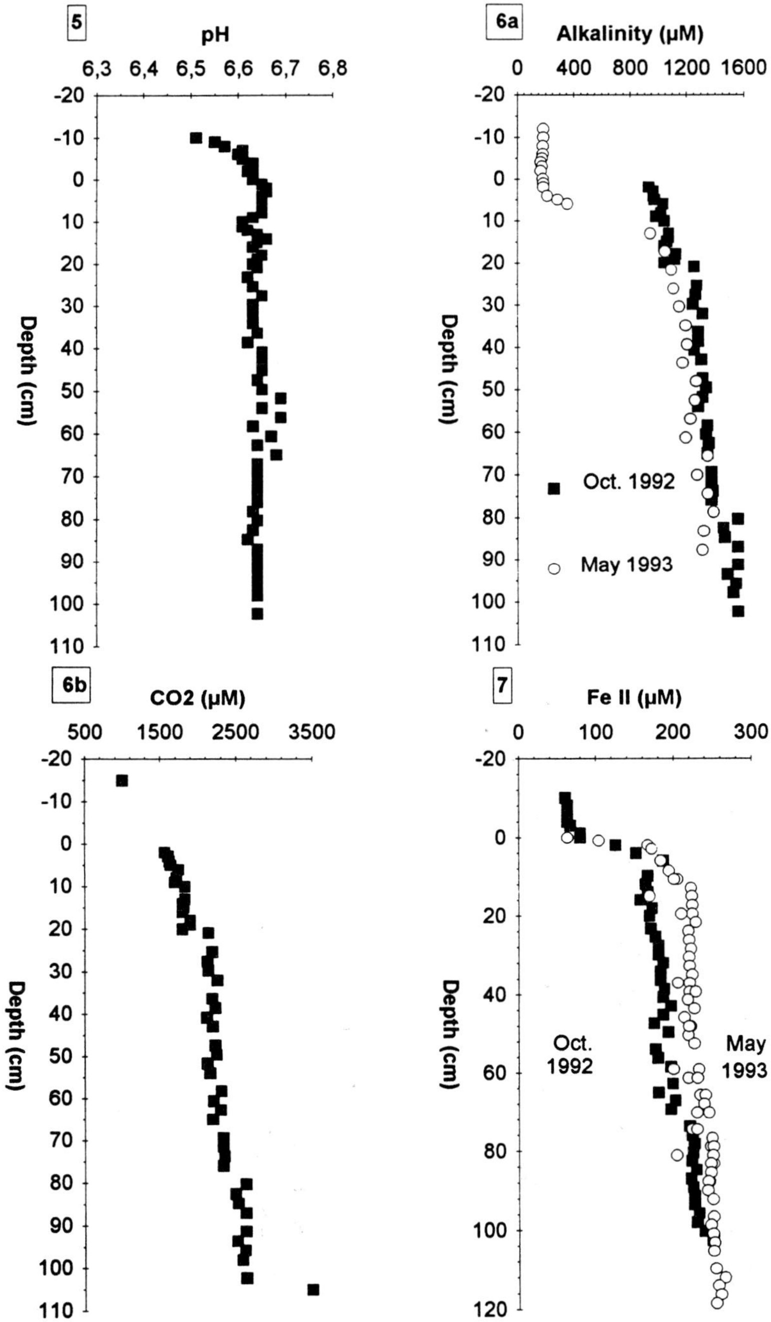

Figs. 5 - 7: 5 - pH pore water (October 1992). 6a - Alkalinity (October 1992 and May 1993). 6b - Total CO_2 calculated from pH, alkalinity and phosphate values : $[CO_2] = (alk - [PO_4]/2)(1 + 10^{(6.49 - pH)})$. 7 - Fe (II).

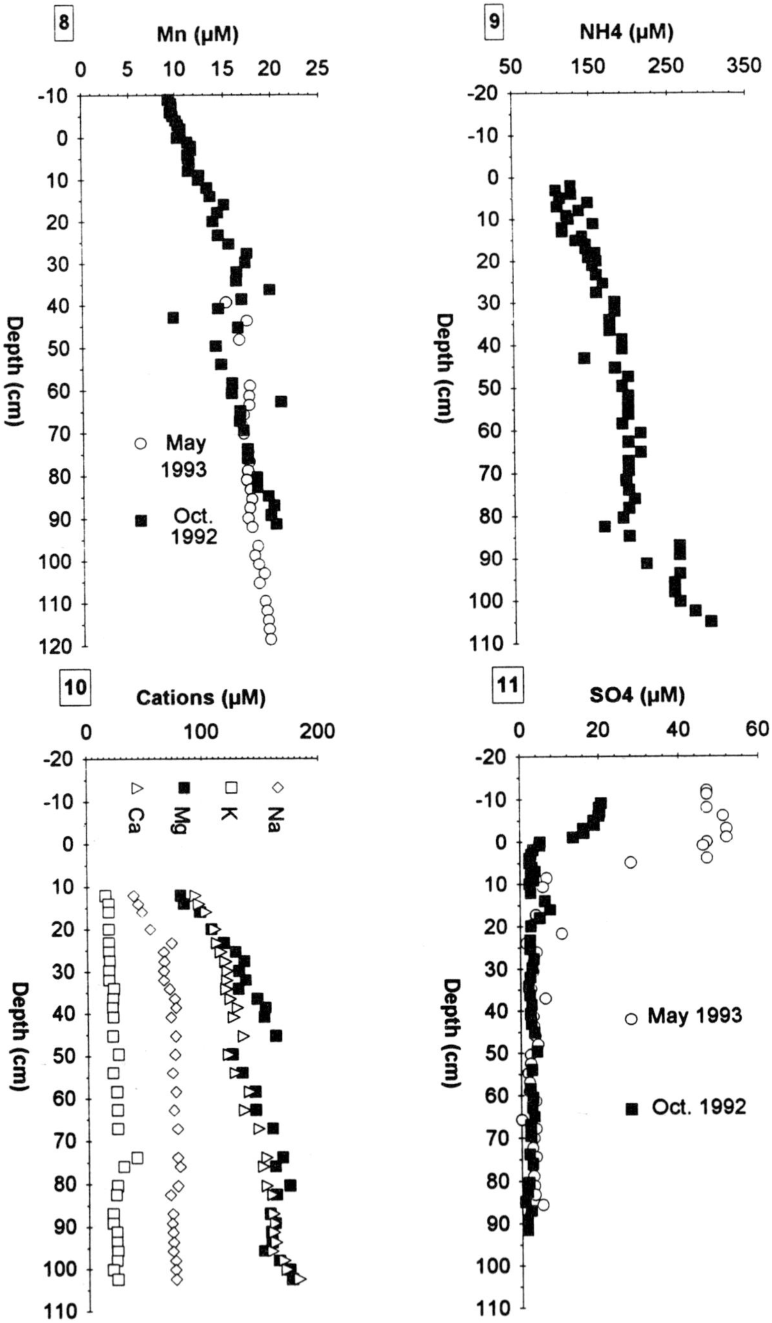

Figs. 8 - 11: 8 - Mn (Oct. 1992 and May 1993). 9 - NH_4^+ (Oct. 1992). 10 - Alkalines and earth-alkalines (oct. 1992). 11 - Sulphate.

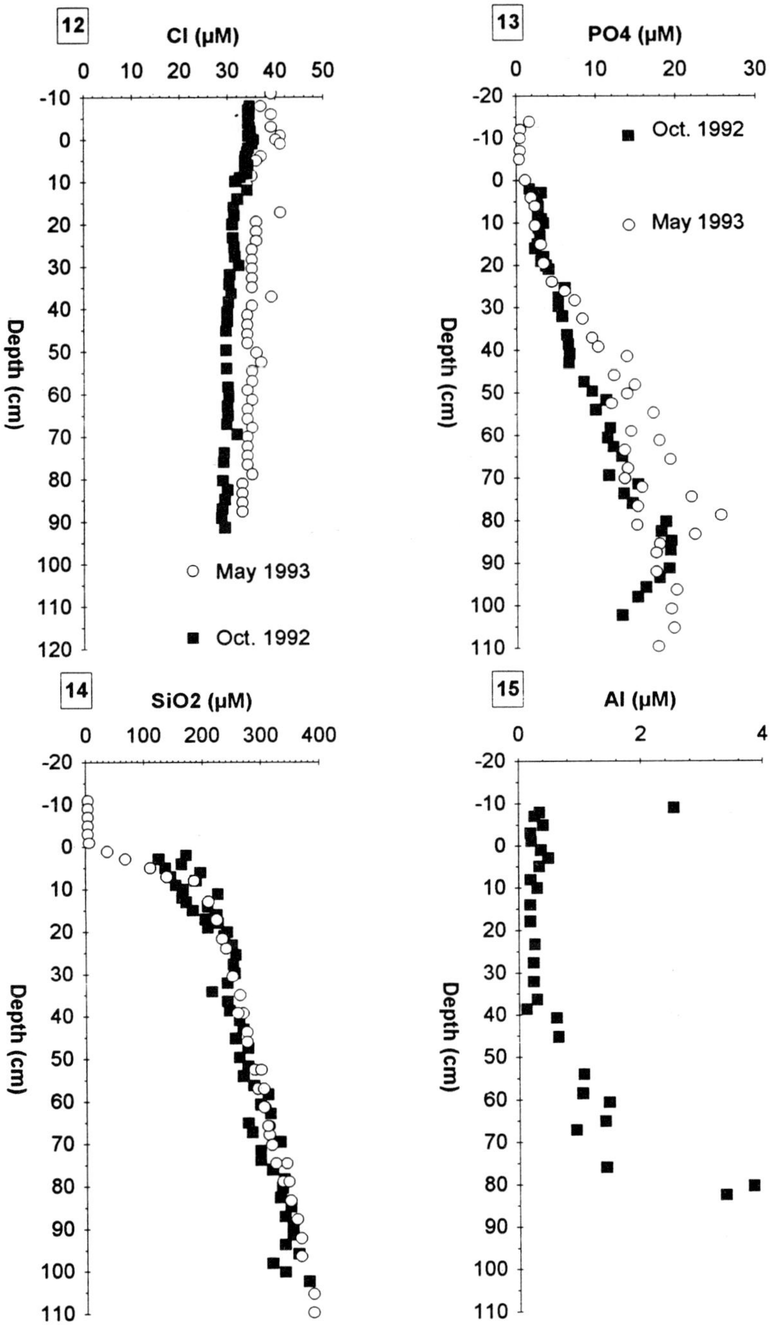

Figs. 12 - 15 : 12 - Chloride. 13 - Orthophosphates. 14 - Dissolved silica in October 1992 and May 1993. 15 - Aluminium in Oct. 1992.

Selected gradients :

Two concentration variation fields can be distinguished from these sections: a strong increase (for iron, ammonia for example) or decrease (for sulphate) occurring in the first centimetres, then below a domain with little variations. Mineralisation reactions are dominating and quick in the first zone near the sediment-water limit, but slower in the second zone. Both zones did not take place exactly at the same depth in spring and autumn, corresponding to different oxygenation rates of the hypolimnion. In October 1992 (hypolimnion base was anoxic), stronger gradients took place from 9 cm above the sediment to 4 cm under the SWI. In May 1993, when the lake water is homogeneous and oxygenated, gradients were situated a little lower, from 0 to 9 cm under the sediment-water limit.

For both seasons, mineralisation processes were virtually equal and show (fig. 16a and 16b) that:

- methanogenesis predominates (90% for the upper section, 97% for the deeper section);
- sulphate reduction represents about 4% of the organic matter degradaion in October and 7% in May (near zero in October and about 1% in May). The difference between both seasons is the most important for this compound, sulphate being partly consumed in the hypolimnion in October.
- contribution of iron reduction is about from 3% in the top section to 2% in the low section.
- contribution of MnO_2 reduction is negligible (1% maximum in the deeper section).

Organic matter formula -

The previous model is used with Redfield formula for OM It is possible to specify stoechiometric coefficients x, y, z corresponding to $(CH_2O)_x(NH_3)_y(H_3PO_4)_z$

Contribution coefficients are ascribed to the main reactions :

- sulphate reduction:

α: $(OM + x/2\ SO_4^{2-} + x/2\ Fe^{2+} \longrightarrow x\ CO_2 + y\ NH_3 + z\ H_3PO_4 + x/2\ FeS + x\ H_2O)$

- iron III reduction :

β: $(OM + 4x\ Fe(OH)_3 + 8x\ H^+ \longrightarrow x\ CO_2 + y\ NH_3 + z\ H_3PO_4 + 11x\ H_2O + 4x\ Fe^{2+})$

- methanogenesis :

$(1-\alpha-\beta)$: $(OM \longrightarrow x/2\ CO_2 + x/2\ CH_4 + y\ NH_3 + z\ H_3PO_4)$

Produced amounts of CO_2 and NH_4^+ can be expressed as :

$$(\Sigma CO_2/\Sigma NH_4) = (\alpha x + \beta x + (1-\alpha-\beta)x/2) / (\alpha y + \beta y + (1-\alpha-\beta)y)$$

so:

$$(\Sigma CO_2/\Sigma NH_4) = (x / 2y)(1+\alpha+\beta)$$

According to Berner (1980), if adsorption is neglected, we have:

$$(\Delta\Sigma CO_2 / \Delta\Sigma NH_4) = (x / 2y)(1+\alpha+\beta)(D_{NH4} / D_{CO2})$$

where D_i is the apparent diffusion coefficient of i ($D_{NH4} = 285$ cm^2/year and D_{CO2} can be considered as D_{HCO3^-} *i.e.* 177 cm^2/year). α and β values are determined in the previous model (respectively 0.06 and 0.04 for upper zone, and about 0 and 0.03 for the deeper one). Gradient ratios for CO_2 and NH_4^+ can be estimated from total mineral carbon versus ammonia (figure 17).

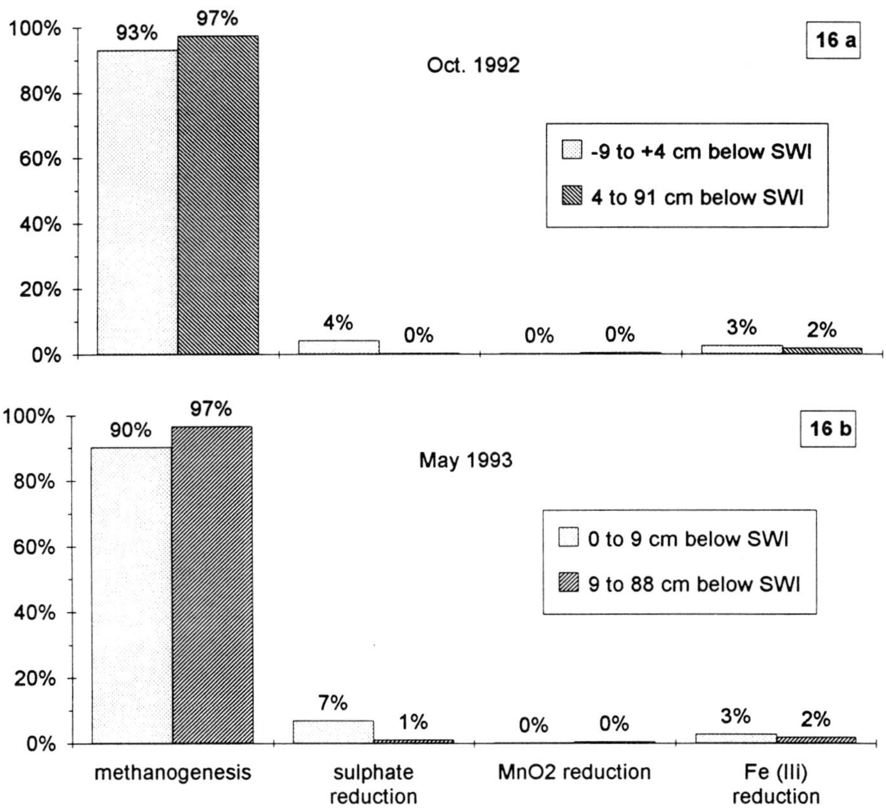

Fig. 16: Proportion of organic matter mineralization process for y = 13.6. a - October 1992. b - May 1993.

So : $x/y = C/N = 6.5*2*0.621/1.03 = 7.8$ if only the deeper (and larger) section is taken into account (the slope of $C=f(N)$ is calculated for this section), and thus $y = 13.6$ if $x = 106$.

The same calculation can be done for x/z ratio :

$$(\Delta\Sigma CO_2 / \Delta\Sigma PO_4) = (x / 2z)(1+\alpha+\beta)(D_{PO4} / D_{CO2})$$

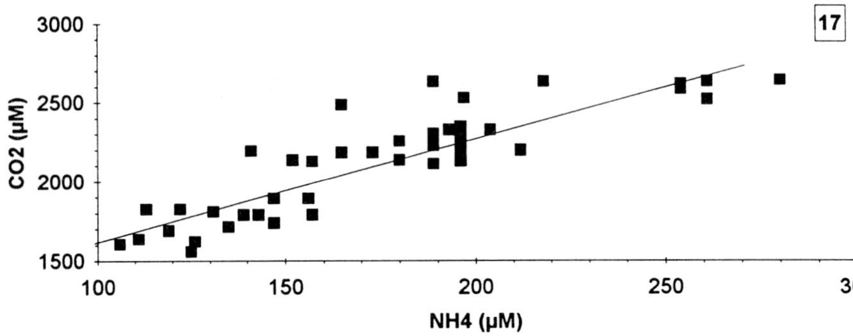

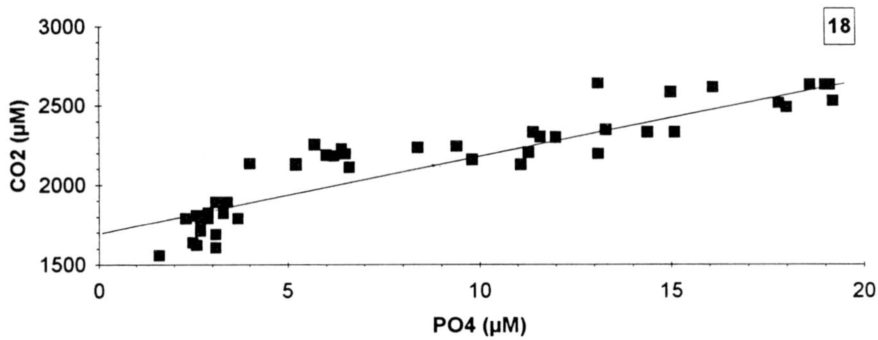

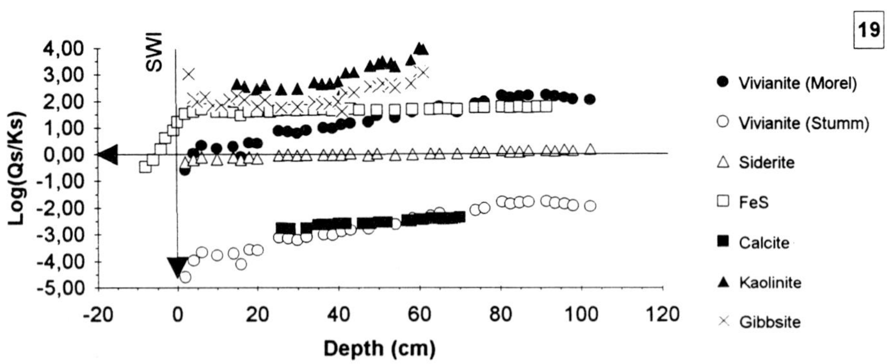

Figs. 17 - 19 : 17 - Total mineral CO_2 versus NH_4^+ (Oct. 1992) : slope = 6.5, r = 0.87. 18 - Total mineral CO_2 versus PO_4 (Oct. 1992) : slope = 50.4, r = 0.91. 19 - Saturating rating for some sompounds.

with D_{PO4} = 138 cm^2/year ($H_2PO_4^-$ and HPO_4^{2-}) and a slope of the representation ΣCO_2 versus PO_4 of 50.4 (figure 18), we find: x/z = C/P = 125.5 so z = 0.84. However, the latter value is given for information only, because of the poorly understood behaviour of phosphorus.

The formula for the organic matter in decomposition in the sediment could be $(CH_2O)_{106}(NH_3)_{13.6}(H_3PO_4)_{0.84}$ These coefficients are smaller than the Redfield ones and are in accordance with other temperate lakes (Hecky, 1993). This formula could correspond to an organic matter of detrital origin rather than of phytoplanktonic origin. On the other hand, the OM degradation begins in the free water of the lake (Viollier *et al.*, this volume), and its composition could be changed if C-N bonds are more rapidly destroyed as suggested by Philippe (1989). It is noteworthy that N/P ratio = 16.2 (near the Redfield ratio).

With this new OM formula, contributions for mineralisation reactions can be recalculated, but the difference with the first estimation is rather unsignificant. The single difference is for October 1992 values in the upper zone (93% methanogenesis against 90% previously, and 4% for sulphate reduction against 6%).

Other compounds-

Some other concentration sections have been determined for compounds which do not directly participate in OM degradation. Silica, alkalines and earth-alkalines, aluminium participate in early diagenesis of sediment.

a) Silica section (figure 14).

Silica concentration versus depth increased irregularly. The main part may come from Diatom skeletons dissolving, identified in the solid phase, and another part from detrital minerals:

$$SiO_2 \text{ solid} + H_2O \rightleftharpoons H_4SiO_4$$

where dissolved silica is mainly on neutral form (pK_a = 9.82 at 298 K).
Silica concentration was low, in agreement with the low Diatom fraction in the sediment. Saturation was not reached (bioopal solubility is about 10^{-3} mol/l). By comparison with the Lac d'Aydat, a eutrophic lake in which Diatoms compose 80% of the sediments, silica saturation is reached.

b) Alkalines and earth-alkalines (figure 10)

Several hypotheses may explain the increase of the concentration observed for these elements :
- release from organic matter (but concentration sections do not correspond to the ammonia one where the stronger increase took place in the first centimetres),
- desorption from iron hydroxides and manganese oxides, consumed in OM degradation,
- ion exchange: adsorbed alkalines and earth-alkalines on clays could be released in solution by exchange with Fe^{2+} or NH_4^+,
- reorganisation of some minerals, releasing a greater part of exchangeable ions,
- mineral dissolution.

In order to answer these questions, exchangeable elements of solid fraction have been determined. Exchangeable potassium (fig. 20) increases with depth, as does dissolved K. The hypothesis of ion exchange does not seem to correspond because the ratio (exchange fraction)/(solution fraction) was constant (about 100). A reorganisation of solid phase, giving more exchangeable cations, could explain the K section. The same interpretation can be advanced for Ca and Mg sections (fig. 21 and 22).

c) Aluminium (figure 15)

Strong concentration variations were determined for aluminium, near a depth of 80 cm in the sediment. Other parameters were also changing at this depth: porosity, total organic carbon, granulometry, phosphate and alkalinity sections, and also iron and ammonia sections for which the variation was not as important as others. Oversaturation of Al can be calculated for gibbsite and kaolinite (fig. 22), but Al analysis must be used with care (sample contamination, diffusion of colloidal aluminium through the membrane were possible).

d) Saturation rating for some compounds (figure 19)

Some compounds released in OM mineralisation may precipitate. Interstitial water was undersaturated for calcite, just about saturated for siderite, and oversaturated for FeS (inaccuracy for thermodynamic data about vivianite does not allow to conclude for this mineral). Minerals like pyrite were in fact identified in superficial sediment, and vivianite and siderite deeper (Truze E., 1990).

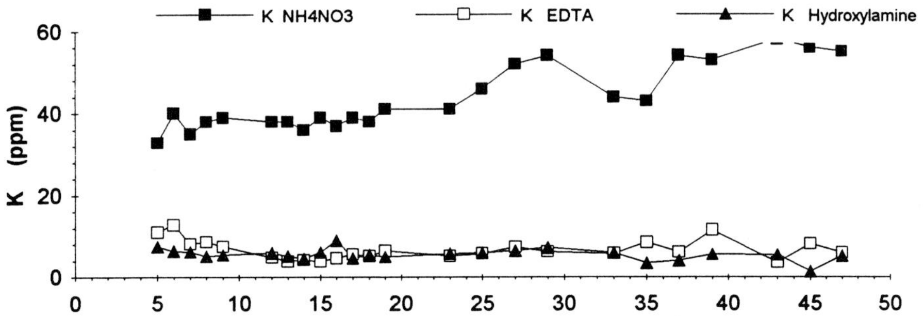

20

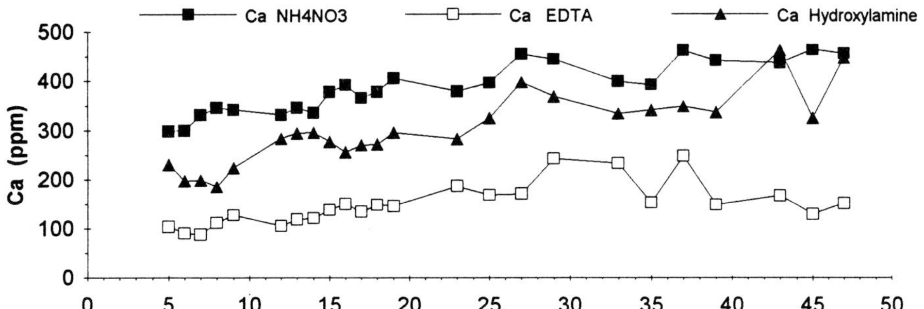

21

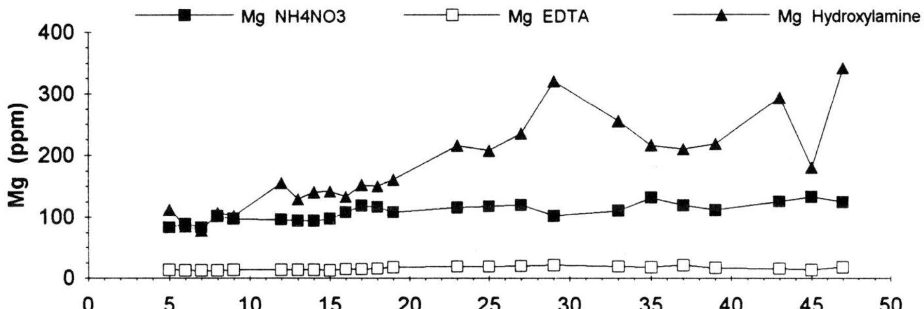

22

Figs. 20 - 22 : 20 - Exchangeable K from solid sediment after 3 successive attacks (1 - NH_4NO_3, 2- EDTA, 3- Hydroxylamine). 21 - Exchangeable Ca. 22 - Exchangeable Mg.

Conclusion-

The Lac du Bouchet presents a very low sedimentation rate for a lacustrine environment. From a pore water composition point of view, the diffusion phenomenon becomes predominant over initial deposition of different materials (sedimentation disparities tend to be cancelled out). So concentration gradients observed are due to current diagenesis reactions in the sediment.

In general, pore water of the Lac du Bouchet are very diluted (following the free water of the lake, which is essentially constituted of rain water). Two domains can be distinguished about organic matter mineralisation. The first part is localised near the sediment-water interface : degradation reactions are fast, producing strong concentration gradients up to 10 to 20 cm (Fe^{2+}, SO_4^{2-}, NH_4^+,Mn^{2+},CO_2). Below this first zone, gradients are lower but OM mineralisation continues, with a 3 to 4 times slower kinetic rate.

In a general way, equilibrium concentration is not reached for any compound. Mineralisation phenomena are roughly in accordance with classical early diagenesis processes. However, some elements have a complex behaviour, like alkaline and earth-alkaline. A reorganisation of minerals and ion exchange (with Fe^{2+} or NH_4^+) could explain these sections. The example of phosphate seems to be an original case of mineralisation, its production in aqueous phase being delayed in comparison with ammonia. It could be due to a ferrous phosphate (vivianite) precipitation, but saturation rating does not allow any firm conclusions on this point.

A formula for the organic matter in decomposition could correspond to $(CH_2O)_{106}(NH_3)_{13.6}(H_3PO_4)_{0.84}$. Nitrogen and phosphorus coefficients are slightly inferior to Redfield stoechiometry. In order to determine true C/N and C/P ratios for total organic matter, particle phase analyses should be carried out. Organic matter transformation is very slow at the Lac du Bouchet and seems to continue up to 2500 years of burying. Another hypothesis for the non stationary state should imply a sediment washing by water infiltration (Truze E., 1990). These losses may be in accordance with water balance of the lake (Viollier *et al.*, this volume), and could explain the oligotrophic state of the old Lac du Bouchet.

Acknowledgements- A special thank is given to Bernard Guillet for his advices and suggestions concerning the sequential chemical attacks of the sediment and to L. Dever (Laboratoire de Géochimie isotopique de l'Université Paris-Sud) for the ^{14}C data. This research was partly financed by ANDRA, and we particularly thank Th. Merceron for his support. We greatly thank the scuba divers of the first-aid post of Le Puy-en-Velay for their invaluable assistance.

References-

Bonifay E. and Truze E. (1987) Dynamique sédimentaire et évolution des lacs de maars: l'exemple du Velay, *In* Documents du C.E.R.L.A.T. n° 1, E. BONIFAY (ed.), pp. 29-64.

Bonifay E. and Truze E. (1991) Histoire géologique du Lac du Bouchet, *In* "Le Lac du Bouchet: environnement naturel et étude des sédiments du dernier cycle climatique". Documents du C.E.R.L.A.T. n° 2, E. BONIFAY (ed.), pp.35-61

Bertrand P., Brocero S., Lallier-Vergès E., Tribovillard N. and Bonifay E. (1992) Sédimentation organique lacustre et paléoclimats du Pléistocène aux moyennes latitudes : exemple du Lac du Bouchet, Haute Loire, France (résultats préliminaires). *Bull. Soc. Géol. France*, **163**(4), 427-433

Carignan R. (1984) Interstitial water sampling by dialysis : methodological notes. *Limn. Ocean.* **29**, 667-670

Morel F.M.M. (1983) Principles of aquatic chemistry, Wiley Intersciences, 186 p.

Hecky R.E., Campbell P. and Hendzel L.L. (1993) : The stoechiometry of carbon, nitrogen, and phosphorus in particulate matter of lakes and oceans. *Limn. Ocean.*, 38, 709-724

Lallier-Vergès E., Sifeddine A, de Beaulieu J-L., Reille M., Tribovillard N.P., Bertrand Ph., Montgenot Th., Thouveny N., Disnar J-R. and Guillet B. (1993) Sensibilité de la sédimentation organique aux variations climatiques du Tardi-Würm et de l'Holocène ; le lac du Bouchet (Haute-Loire, France). *Bull. Soc. Géol. France*, **164**(5), 661-673

Patience A., Lallier-Vergès E., Albéric P., Desprairies A. and Tribovillard N.P. (1994) Relationships between organo-mineral supply and early diagenesis in the lacustrine environment : a study of surficial sediments from the Lac du Bouchet (Haute-Loire, France), *Quaternary Sci. Rev.* (in press)

Philippe L. (1989) Bilan géochimique du fer et du phosphore dans un écosystème lacustre eutrophe : le lac d'Aydat (Puy de Dôme) Thèse Université Paris VII, 186 p.

Podda F. and Michard G. (1994) Mesure colorimétrique de l'alcalinité. *C. R. Acad. Sci. Paris*, in press.

Stumm W. and Morgan J.J. (1981) Aquatic chemistry, Wiley Interscience, 245 p.

Thurman E.M. and Malcolm R.L. (1981) Preparative isolation of aquatic humic substances. *Environm. Sci. Technol.*, **15**, 463-466.

Truze E. (1990) Etude sédimentologique et géochimique des dépôts du maar du Bouchet. Evolution d'un système lacustre au cours du dernier cycle climatique (0-120 000 ans BP). Thèse Université Aix-Marseille II. 242 p.

Table 1 - Pore water composition, October 1992. Concentrations in 10^{-6} mol/l.

prof.	P1	P2	P3	pH	Alc	Alc* (calc.)	RB	SCO2	C tot	Na	K	Mg
-10		2		6,51								
-9	1	3		6,55								
-8	2	4		6,57								
-7	3	5		6,61								
-6	4	6		6,60								
-5	5	7		6,61								
-4	6	8		6,63								
-3	7	9		6,63								
-2	8	10		6,62								
-1	9	11		6,63								
0	10	12		6,63								
1	11	13		6,65								
2	12	14	1	6,66	930			1557	2959			
3	13	15	2	6,66	960			1606	3052			
4	14	16	3	6,65	960			1622	3082			
5	15	17	4	6,65	970			1639	3114			
6	16	18	5	6,65	1030			1740	3307			
7	17	19	6	6,65					0			
8	18	20	7	6,65	1015			1715	3258			
9	19		8	6,63	980			1687	3206			
10	20	21	9	6,61	1040			1826	3469			
11			10	6,61					0			
12	21	22	11	6,62		838,5	724,5		0	41	16	82
13			12	6,64	1070			1825	3468			
14	22	23	13	6,66	1070	880,6	744,5	1791	3403	45	19	85
15			14	6,64	1060			1808	3436			
16	23	24	15	6,63	1040	909,7	769	1791	3404	48	19	98
17			16						0			
18	24	25	17	6,65	1120			1892	3595			
19			18	6,64	1110			1893	3597			
20	25	26	19	6,63	1040	993,6	840,3	1790	3401	55	19	108
21			20	6,64	1250			2132	4050			
23	26	27	21	6,62		1045,8	888,8		0	74	19	118

Table 1 (*continued*).

prof.	Ca	Fe(II)	Mn	Al	NH4	Cl	SO4	PO4	SiO2	S+	S-	dif (%)	I
-10		60											
-9			9,3	2,54		47,9	20,6						
-8		63	9,6	0,34		34,6	20,1						
-7			9,7	0,26		34,4	20,3						
-6		63	9,5			34,6	19,8						
-5			9,8	0,40		34,4	18,6						
-4		63	10,1			34,4	18,8						
-3		67	10,3	0,20		34,6	16						
-2			10,6			34,8	16,2						
-1		80	10,6	0,21		34,4	13,3						
0		79	10,2			35,4	5,1						
1			11,3	0,37		35,1	4,8						
2		125	11,6		125	34,4	3,3	1,6	172	398,2	972,6	84	0,000825828
3			11,6	0,48	106	34,1	2,8	3,1	126	129,2	1002,8	154	0,000581423
4		151	11,3		126	33,8	2,2	2,6	164	450,6	1000,8	76	0,000891058
5			11,4	0,33	111	33,7	2,3	2,5	137	133,8	1010,8	153	0,000586825
6		186	11,5		147	34,3	2,9	2,7	197	542,0	1072,8	66	0,001008691
7					108	33,7	3,8		145	108,0	41,3	89	0,00007845
8			11,4	0,20	135	34,1	2,6	2,7	189	157,8	1057,0	148	0,000622291
9			12,5		119	32,5	3,4	3,1	154	144,0	1022,4	151	0,000600123
10		167	12,4	0,30	122	31,7	2,2	3,3	167	480,8	1079,4	77	0,000962789
11					153			2,8	226	153,0	2,8	193	0,000078824
12	95	164	13,3		114	34,1	2,6		165	877,8	39,3	183	0,00081475
13					113			2,9	172	113,0	1072,9	162	0,000593907
14	98	167	13,6	0,18	139	32,1	6,2	2,9	208	928,0	1117,4	19	0,001392657
15					131			2,6	183	131,0	1062,6	156	0,000597658
16	104	157	15		143	31,2	7,4	2,3	224	958,0	1088,3	13	0,001405609
17					143				204	143,0	0,0	200	0,0000715
18		172	14,3	0,18	156	31,3	5	3,4	226	528,6	1164,7	75	0,001039072
19					147			3,1	208	147,0	1113,1	153	0,000631073
20	111	169	13,9		157	30,9	2,6	3,7	242	1033,4	1079,8	4	0,001461921
21					152			4,0	237	152,0	1254,0	157	0,00070432
23	113	171	14,4	0,26	157	31,1	2,5		249	1081,9	36,1	187	0,0009777

Table 1 (*continued*).

prof.	Ca	Fe(II)	Mn	Al	NH4	Cl	SO4	PO4	SiO2	S+	S-	dif (%)	I
25	116	177	15,5		165	31,4	2,5	6,0	256	1124,7	1312,4	15	0,00166003
28	120	181	17,4	0,24	157	31,5	3,4	5,2	252	1149,7	1303,5	13	0,001684916
30	122	181	17,2		180	32,3	3	5,2	254	1167,6	1283,5	9	0,001681166
32	121	186	16,3	0,24	180	30,4	2,5	5,7	242	1184,9	1351,1	13	0,001732181
34	121	183	16,2		173	30,3	2		215	1167,2	34,3	189	0,00105285
36	124	183	19,7	0,29	173	30,7	2,4	6,2	242	1217,1	1321,7	8	0,001746946
39	130	188	16,8	0,12	189	30,2	2,7	6,4	245	1007,8	1322,0	27	0,001787212
41	126	186	14,3	0,61	189	29,8	2,5	6,6	262	1017,4	1291,4	24	0,001750078
43		196	9,7		141	29,8	2,6	6,5	269	1023,4	1341,5	27	0,001157395
45	135	186	16,4	0,64	180	29,6	3,4		255	584,8	36,4	177	0,00116015
47		175			196			8,4	276	1056,6	1318,4	22	0,001109972
50	122	193	14,1		189	29,6	4,1	9,4	262	1157,3	1377,2	17	0,001750852
52					196			11,1	276	770,7	1321,1	53	0,000762213
54	128	177	14,6	1,05	196	29,7	2,8	9,8	269	1175,1	1325,1	12	0,001719234
56		180			196				286	1147,2	0,0	200	0,000458
58	140	197	15,8	1,03	189	30	2,3	11,6	310	1214,9	1386,2	13	0,001837728
61			15,7	1,47	212	30,1	2,9	11,3	297	839,2	1377,2	49	0,000832629
63	136	199	20,9		196	29,9	2,8	12,0	314	1274,5	1397,5	9	0,00185491
65		181	16,6	1,40	212	30	3,3	13,1	276	1273,8	1389,7	9	0,001203673
67	147	202	16,5	0,93	196	29,7	2,4		283	1284,8	34,5	190	0,00121895
69		196	16,9		196	31,9	2,3	11,4	331	621,8	1417,9	78	0,001238812
72					193			15,1	297	790,7	1385,1	55	0,000794033
74	154	221	17,3		196	29,2	2	13,3	297	1341,6	1426,5	6	0,001997789
76	152	224	17,3	1,42	204	29,1	2,7	14,4	317	1348,4	1418,9	5	0,001982602
78		226			196				337	1363,8	0,0	200	0,00055
80	154	225	18,3	3,84	189	29	1,8	18,6	334	1411,6	1601,2	13	0,002096238
83	159	223	18,3	3,36	165	29,8	1,4	18,0	331	1410,9	1500,6	6	0,00201139
85		229	19,5		197	29,3	0,8	19,2	348	1451,8	1510,1	4	0,001357686
87	160	222	20,1		261	28,9	2,3	19,1	338	745,2	1602,6	73	0,002104003
89	161	225	19,7		261	28,6	1,3		351	1486,5	31,2	192	0,0013277
91	161	226	20,3		218	29,3	1,3	19,0	351	1448,2	1600,9	10	0,00209392
94	162	226			261			17,8	338	1451,0	1497,8	3	0,002026274
96	159	232			254			16,1	362	1437,2	1556,1	8	0,002044663
98	169	230			254			15,0	317	1481,3	1535,0	4	0,0020768
100	171	240			261				338	1526,6	0,0	200	0,0013477
102	182	250			280			13,1	379	1595,1	1563,1	2	0,002190423

Table 1 (*continued*).

prof.	P1	P2	P3	pH	Alc	Alc*	RB	SCO2	C tot	Na	K	Mg
25	27	28	22	6,63	1270	1082,3	923,3	2185	4151	67	19	128
28	28	29	23	6,65	1260	1106,2	954,4	2127	4042	67	19	135
30	29	30	24	6,63	1240	1124,1	949,3	2134	4054	67	19	131
32	30	31	25	6,63	1310	1143,8	969,5	2254	4283	67	19	137
34	31	32	26	6,63		1132,9	959,9		0	72	22	130
36	32	33	27	6,64	1280	1175,4	1008,6	2181	4144	76	22	147
39	33	34	28	6,62	1280	1223,2	1040,6	2223	4224	77	22	154
41	34	35	29	6,65	1250	1200	1017,6	2109	4007	73	22	152
43	35	36	30	6,65	1300			2194	4168			
45	36	37	31	6,65		1241,3	1061,3		0	77	21	163
47		38	32	6,64	1310			2230	4237			
50	37	39	33	6,65	1330	1153,7	974,1	2242	4260	76	26	125
52		40	34	6,69	1310			2128	4042			
54	38	41	35	6,65	1280	1151,2	965	2157	4099	74	22	133
56		42	36	6,69					0			
58	39	43	37	6,63	1340	1237,4	1060	2301	4371	77	25	144
61	40	44	38	6,67	1330			2199	4179			
63	41	45	39	6,64	1350	1250,1	1066,1	2295	4361	75	25	145
65	42	46	40	6,68	1340			2194	4169			
67	43	47	41	6,64		1314,3	1118,3		0	78	25	159
69	44	48	42	6,64	1370			2330	4427			
72		49	43	6,64	1370			2327	4421			
74	45	50	44	6,64	1380	1389,4	1206,7	2346	4457	78	42	168
76	46	51	45	6,64	1370	1373,7	1184,1	2328	4422	80	30	162
78		52	46	6,63					0			
80	47	53	47	6,64	1550	1382,2	1211,8	2631	5000	78	25	174
83	48	54	48	6,63	1450	1333,1	1186,1	2485	4721	71	24	162
85	49	55	49	6,62	1460			2526	4799			
87	50	56	50	6,64	1550	1419	1177,1	2631	4999	73	21	157
89	51	57	51	6,64		1456,8	1195,8		0	73	21	161
91	52		52	6,64	1550	1392,7	1193,7	2631	4999	74	24	157
94	53		53	6,64	1480			2513	4774	74	24	158
96	54		54	6,64	1540			2616	4971	74	24	152
98	55		55	6,64	1520			2583	4908	76	24	165
100	56		56						0	76	21	173
102	57		57	6,64	1550			2636	5009	76	25	175

Table 2 - Pore water composition, May 1992. Concentrations in 10^{-6} mol/l.

prof	Q1	Q2	Q3	pH	Alc	SCO2	Na	K	Mg	Ca	Fe 1	Fe2	Fe 3	Mn	Cl	SO4	PO4	Si 2	Si 3
				6,65															
-15		1																	
-14		2															1,7		
-13		3																	
-12	1	4			185	313	46	9	63	125					39	47,0	0,5		
-11	2	5													39	47,0		3,3	
-10	3	6			183	309	42	11	74	126							0,5		
-9	4	7																3,3	
-8	5	8			179	303	45	11	70	142					37	47,0			
-7	6	9															0,5	3,3	
-6	7	10			177	299	36	9	64	105					39	51,0			
-5	8	11			172	291											0,3	3,3	
-4	9	12			164	277	42	9	65	148									
-3	10	13			171	289									39	52,0		3,3	
-2	11	14			167	283	34	10	51	78									
-1	12	15													41	52,0		5,4	
0	13	16			180	304	41	9	59	116		63			40	47,0	1,1		
1	14	17			183	310						104			41	46,0		36,5	
2	15	18			183	310	40	11	58	101		166,3							
3	16	19										171,1						68,4	
4	17	20			211	355	48	12	87	176					37	47,0	1,8		
5	18				284	480									36	28,0		111,0	
6	19	21			351	592	43	15	104	141		182,7					2,3		
7	20																	139,0	
8		22																185,0	
8,5	21						47	14	90	171	194				35	6,7			
10,7	22	23									205	200,7			34	5,8	2,3		
12,9	23	24			940	1590	45	16	133	178	222							209,0	
15,1	24	25									223	168,8					3,0		
17,3	25	26			1043	1765					224				41	3,9		224,0	
19,5	26	27									224	209,4			36	2,9	3,3		
21,7	27	28			1090	1844	43	15	147	173	228				36	10,5		233,0	
23,9	28	29									219				36	1,5	4,4	239,0	
26,1	29	30			1105	1864	47	16	139	206	220				35	4,1	6,0		
28,3	30	31									222				35	3,3	7,2		
30,5	31	32			1145	1937	41	19	155	174	220				35	3,3		251,0	
32,7	32	33									221				35	2,5	8,2		

Table 2 (*continued*).

prof	Q1	Q2	Q3	pH	Alc	SCO2	Na	K	Mg	Ca	Fe 1	Fe2	Fe 3	Mn	Cl	SO4	PO4	Si 2	Si 3
34,9	33	34			1190	2013	50	16	152	268	224				35	2,6		263,0	
37,1	34	35									221	205,8			39	6,2	9,4		
39,3	35	36	21		1200	2022	47	23	177	199	221		228	15,2	35	2,7	10,2	269,0	260
41,5	36	37	22								218				34	3,1	13,7		
43,7	37	38	23		1166	1973	53	21	158	249	226		227	17,3	34	3,3		275,0	
45,9	38	39	24								214				34	3,6	12,1	275,0	
48,1	39	40	25		1265	2128	40	21	179	191	222		220	16,5	34	4,2	14,7		
50,3	40	41	26								219				36	2,5	13,8		
52,5	41	42	27		1255	2113	51	22	147	183	226				37	2,5	11,8	287,0	299
54,7	42	43	28												35	1,9	17,1		
56,9	43	44	29		1222	2067	51	23	191	226					35	2,1		293,0	302
59,1	44	45	30									200,4	232	17,6	34	2,0	14,2		
61,3	45	46	31		1190	1998	56	21	156	232	219		231	17,5	35	3,5	17,8	305,0	302
63,5	46	47	32											17,5	34	2,8	13,4		
65,7	47	48	33		1340	2251	56	24	169	227	234		241	16,9	34	0,0	19,1	311,0	309
67,9	48	49	34										239	16,7	35	3,5	13,8	311,0	
70,1	49	50	35		1270	2137	56	24	169	227	230		245	16,9	34	3,1	13,4		316
72,3	50	51	36												34	2,7	15,6		
74,5	51	52	37		1340	2249	54	25	206	225	224		230	17,4	34	3,6	21,7	323,0	341
76,7	52	53	38										249	17,5	34	2,7	15,0		
78,9	53	54	39		1382	2317	54	22	165	181	248		251	17,3	35	3,0	25,4	335,0	345
81,1	54	55	40									203,9	250	17,2	33	3,1	14,9		
83,3	55	56	41		1317	2209	50	22	204	222	251		248	17,6	33	3,3	22,2	347,0	348
85,5	56	57	42										248	17,7	33	5,2	17,8		
87,7	57		43		1305	2193	62	26	186	281	246		244	17,5	33	2,0	17,4		359
89,9			44										243	17,3					
92,1			45										250	17,7			17,4		366
94,3			46																
96,5			47										251	18,3			19,9		366
98,7			48										248	18,0					
100,9			49										250	18,4			19,2		
103,1			50										252	19,0					
105,3			51										251	18,5			19,6		387
107,5			52																
109,7			53										253	19,1			17,6		387
111,9			54										265	19,3					
114,1			55										257	19,4			16,7		394
116,3			56										260	19,5					
118,5			57										254	19,6			12,6		

Organic fluxes and early diagenesis in the lacustrine environment: the superficial sediments of the Lac du Bouchet (Haute Loire, France)

Andrew J. Patience, Elisabeth Lallier-Vergès, Abdelfettah Sifeddine, Patrick Albéric and Bernard Guillet.

Université d'Orléans, URA 724 du CNRS, Dépt. des Sciences de la Terre, F-45067 Orléans cedex

Key-words- *Lac du Bouchet, organic matter, early diagenesis, methanogenesis, organic fluxes.*

Abstract- Superficial sediments from the Lac du Bouchet were analysed to study the degradation processes which affect sedimentary organic matter in an oligotrophic lacustrine environment. This was undertaken by the study of the evolution of the organic matter composition, in terms of both early diagenesis and any possible variations in organic inputs. The petrographical study of the resistant organic matter shows that some marked variations in the type, nature and abundance of organic inputs have occurred through time over the last 2500 years. The geochemical study of the bulk organic matter and the distribution of alkali-soluble components indicate that the effects of early diagenesis are only visible when the organic inputs are unchanged (< 1400 years). Due to the very low sedimentation rates, methanogenesis seems only to affect the autochthonous (algal, phytoplanktonic) organic matter which consists of a degraded and amorphous organic matter, whereas the organic matter deriving from the surrounding basin has been partially or totally degraded before its deposition.

Introduction-

Recent climato-stratigraphic studies of Late Glacial sediment, from temperate lacustrine environments have illustrated how sedimentary organic material can accurately record climatic variations (Bonifay and Truze, 1987; Bertrand *et al.*, 1992; Lallier-Vergès *et al.*, 1993; Rein and Negendank, 1993; Meyers and Ishiwatari, 1993).

The heterogeneity of the sedimentary organic material is partly the result of the diversity in the original constituents and partly due to the diverse biogeochemical transformations which occur before deposition and during early diagenesis. This work aims to characterise and measure, with a high resolution, the signature of early diagenesis on the organic composition in the upper 50cm of lacustrine sediments, to evaluate the organic fluxes during the Sub-Atlantic, and finally to study the interaction between these fluxes and early diagenesis processes.

The Lac du Bouchet is a maar crater lake some 28 m deep, situated in the Devès volcanic Massif (15km SW of Le Puy; 44° 55'N, 3° 47'W) at 1205m altitude (see fig. 1 Sifeddine *et al.*, this volume). Maar lakes are found throughout the world but the European ones, being situated in the temperate belt, are most likely to reflect subtle changes in the palaeoclimate. Moreover, their altitude (700-1300 m) makes them particularly susceptible to changes in climate, and may even serve to amplify variations in such systems. Maar lake sedimentation depends essentially on the stability and composition of the water column (critically dependent on the surrounding basaltic basin and rainfall); a well defined and constant catchment basin ensuring a homogeneity of detrital input; and climate change which may trigger changes in vegetation, sedimentation pathways with respect to erosion factors, eolian input and the quality and quantity of organic material (Truze, 1990). The Lac du Bouchet displays relatively constant and sufficiently rapid sedimentation over the past 120 kyrs (Bonifay and Truze, 1987; Truze, 1990) to have recorded European climatic variations (Bonifay *et al.*, 1987). The lake waters have a composition close to that of rainfall (Viollier *et al.*, this volume) and the lake is seasonally oxygenated which contributes to its virtual oligotrophy (Jézéquel *et al.*, this volume). The mineral fraction composition of sediments, which is dominated by the basaltic volcani-clastic debris, has not changed significantly over the past 2500 years and no carbonate is present in the sediments (Patience *et al.*, 1994). Thus, the Lac du Bouchet represents an excellent example to observe the diagenetic signature on the organic composition in lacustrine sediments.

Sampling and Methods-

Four sediment cores (S1 to S4) were recovered by divers (in October 1992) by insertion of PVC tubes (each approximately 80 cm in length) into the sediment, which allowed good recovery and preservation of the sediment-water interface. These cores were recovered close to each other in the centre of the lake. Once ashore, the supernatant water was removed by syringe and the cores were flushed with N_2 (to retard any further oxidation of the surface sediments) before being transported upright to the laboratory in an isothermal container. There, cores were sliced in half lengthways and logged for changes in texture, colour and grain size. Subsequently, the cores were sampled (centimetric) after the uppermost three centimetres of sediment have been removed by syringe due to their extremely high water contents. Pore waters were sampled and analysed using dialysers (see Jézéquel *et al.*, this volume).

Rock eval pyrolysis (total organic carbon content and hydrogen index) was performed on total sediment samples (every centimetre) in each core and on the humine fraction from certain levels in core S4. The oxygen index was analysed only on sediments from core S3. Analysis of the palynofacies (optical study of the isolated organic material in the sediment after HCl-HF attack) was performed every 2 cm in core S3. Major element analysis of the isolated organic material was completed on ten samples throughout core

S1. The separation and elemental analysis of the humic substances by titration was completed every 5 cm in core S4. Sulphur was determined on the solutions of ten samples from core S1 following NaOBr attack.

The chronostratigraphy was realised by ^{14}C dating (without $\partial^{13}C$ correction) on four total sediment samples throughout core S3. Specific organic constituents were prepared following a procedure developed by Boussafir *et al.* (1994) in order to be studied by transmission electron microscopy (TEM). These constituents were recovered, from the isolated organic matter, using a micro-manipulation system incorporating a stereo-microscope and syringe, then fixed in osmic acid and set in resin before being cut into ultra-thin sections ready for TEM analysis.

Results and Discussion-

Organic matter abundance-

The sediments from the three cores were found to be almost identical lithologically. Two progressive changes in the colour and texture of the sediment allowed the definition (from the top to the bottom of the cores) of three units: IA, IB and II (fig. 1a). Units I and II have already been defined in cores from the centre of the lake (Sifeddine *et al.*, 1992; Lallier-Vergès *et al.*, 1993). The sediment density increases with depth in unit I due to the progressive compaction during sedimentation, and decreases markedly in unit II due to the facies (lithological) change (fig. 1a).

These lithological changes are also visible in the profile of total organic carbon (TOC) which decreases from around 11% at the surface to approximately 5% at the base of unit IA then remains relatively constant during unit IB before increasing markedly in unit II to values in excess of those at the surface (fig. 1c).

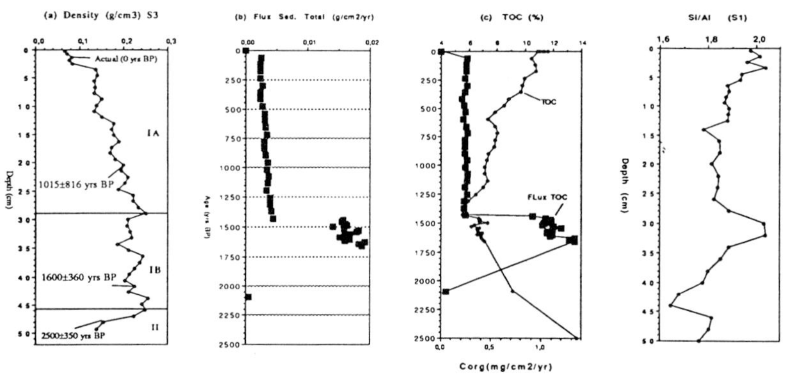

Figure 1. a: variations in sediment density with depth (cm); b: variations in the total sediment flux with age (years BP); c: variations in the values of TOC and organic carbon flux with age; d: profile of the Si/Al ratio in the total sediment with depth (cm).

Genetic origin of the organic matter-

Petrographical studies were performed on the resistant organic matter. Both the organic matter isolated from the mineral groundmass by HCl-HF attack and the non alkali-soluble organic matter (humine fraction) were examined under optical microscope.

Seven main petrographical groups have been observed in the organic matter isolated from the mineral phases by acidic attacks. The dominant type is a greyish amorphous organic material (MOA-G in fig. 2a; G: Pl. 1a). Ultra-thin sections of this material, examined by TEM, revealed an organic matter essentially composed of laminar structures (Pl. 2) corresponding to the resistant outer walls of micro-algae (Derenne *et al* , 1992a, 1992b).

Moreover, this greyish amorphous organic matter is totally absent in the smear-slides prepared from the soils of the surrounding basin. From these observations, the G-AOM is considered to derive from organic phytoplankton or micro-algae and thus is thought to be autochthonous and representative of the sedimentation basin.

The other constituents represent the resistant debris of higher plants and are considered as allochthonous. Their resistant character is either inherited (cuticles [CM: Pl. 1a, c]; spores and pollen [SP: Pl. 1f, g] and preserved ligno-cellulosic material [LCT: Pl. 1d]), or acquired during degradation in the surrounding basin (oxidised and carbonised ligno-cellulosic material [LCO: Pl. 1e, Pl.1h] and reddish amorphous organic material [R: Pl. 1b]). The latter was found to be dominant in smear-slides prepared from the soils of the surrounding catchment basin and is therefore defined as pedogenetic. The carbonised organic components (forest fire debris) and the spore-pollens record both local and regional allochthonous input whilst the other allochthonous fractions reflect strictly the input of the catchment basin.

The heavy fraction of the humine material (obtained after alkaline attacks and observed by optical microscopy) is composed of oxidised ligno-cellulosic debris. The light fraction of the humine was shown to have a large proportion of spores, pollen, cuticles, membranes and fungal filaments [Pl. 1f, 1c, 1d, 1l]. These represent the fraction which was humified before sedimentation.

Between 700-1200 years BP, the proportion of higher plant debris (translucid ligno-cellulosic material and/or oxidised ligno-cellulosic material as well as cuticles and membranes) increases whereas the spore/pollen and the pedogenetic organic material (R-AOM) remain in trace quantities (fig. 2b-2f). These results indicate that a vegetal cover was in place in the surrounding basin at this time. The relative abundance of the oxidised ligno-cellulosic material and the small quantities of pedogenetic material suggest a degradation during a cool and dry climate (Lallier-Vergès *et al.*, 1993; Bertrand *et al.*, 1992).

Between 1400-1700 years BP, the autochthonous organic material (G-AOM and fragments of zooplankton, Pl. 1j, 1k) increases markedly. This level is characterised equally by the increased abundance of Diatoms (observed in smear slides) as by an increase in the Si/Al ratio (thought to be a proxy of biogenic silica) in the total sediment (fig. 1d). The authochthonous/allochthonous ratio in organic matter also increases at this level, suggesting an increased primary productivity in the lake perhaps driven by an increased input of nutrients from the surrounding basin due to anthropogenic deforestation as proposed by Sifeddine *et al.* (1992).

Between 2500-1700 years BP, the proportion of the autochthonous organic material decreases markedly, and consequently, the other constituents increase at this point especially the spore/pollen and pedogenetic material (fig. 2e and 2f). This indicates an increase in the organic input to the lake which is dominated by higher plant debris. Moreover, the relative abundance of the pedogenetic debris illustrates a more marked erosion of the surrounding basin.

Degradation of the organic matter-

The variations in the different forms of carbon (DOC, CO_2 et CH_4) in the pore water illustrate that the degradation of the organic material is, at present, dominantly the result of methanogenesis as CH_4 is the dominant form of carbon found in the pore waters (Jézéquel *et al.*, this volume).

Despite the weak concentration of sulphate in the lake waters (Viollier *et al.*, this volume), there are some traces of sulphate reduction in the form of solid sulphur found close to the sediment water interface (fig. 3c). The pyrite framboids observed under light microscopy are always associated with vegetal debris. Parallel studies of the speciation of sulphur in the first two metres of sediment in the lake during the different climatic phases of the Holocene, have shown the importance of the terrestrial inputs of sulphur (biological in origin) in the establishment of a more intense sulphate reduction during certain periods (Guillet and Maman, this volume).

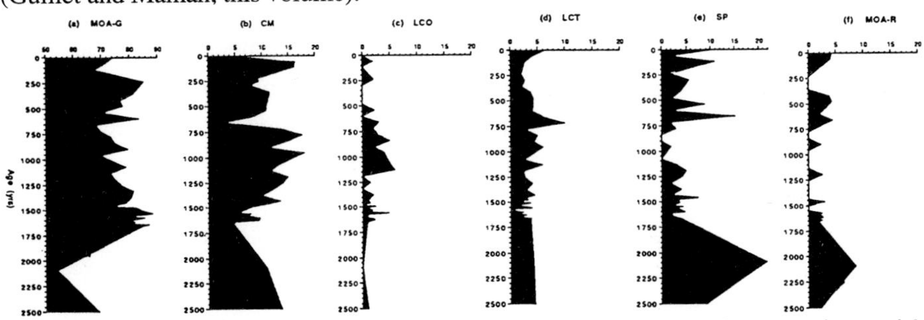

Figure 2. Palynofacies composition (%) over time. a: MOA-G: greyish amorphous organic material (planktonic/autochthonous); b: CM: cuticle membranes; c: LCO: opaque ligno-cellulosic debris (oxidised); d: LCT: translucid ligno-cellulosic debris (preserved); e: SP: spores-pollen and f: MOA-R: reddish amorphous organic material (pedogenetic).

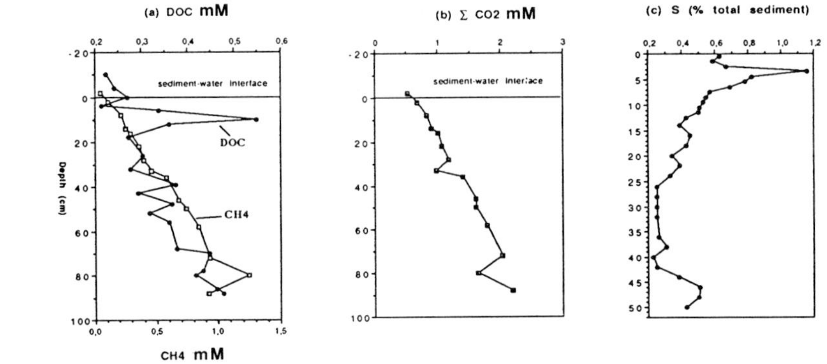

Figure 3. a: profile of the concentration of dissolved organic carbon (DOC) and methane (CH_4) in the pore waters; b: profile of the concentrations of inorganic carbon in the pore waters and c: variations in the solid phase sulphur in the total sediment with depth (cm).

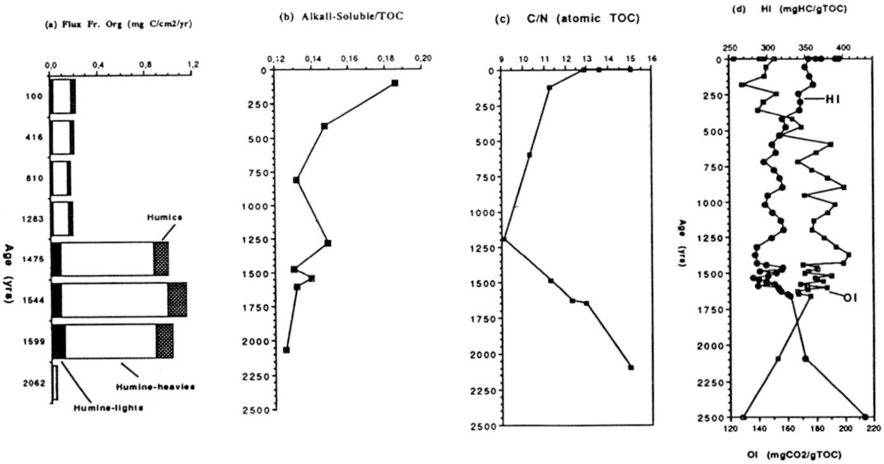

Figure 4. a: variations in the flux of the alkali-soluble humic substances (shaded), in the flux of the humine (light fraction in black, heavies in white); b: proportions of the humine fraction relative to the total organic material; c: variations in the C/N ratio (atomic) of the total organic material with age; d: variations in the HI and OI with age.

The degradation rate of organic carbon has been tentatively modelled, on the basis of the vertical profile of TOC content (fig. 5) and the linear sedimentation rates (calculated between each radiocarbon date). A comparison has been made using data from the sediments of the Lac d'Aydat (Sarazin *et al.*, 1992). This lake is characterised by a very high sedimentation rate in accordance with the eutrophic nature of the lake waters

The results have been superimposed on a diagram (fig. 6), first used by Canfield (1989) in order to compare sulphate reduction and oxic degradation processes in marine environment. Thus, it is shown that the methanogenetic degradation process leads to similar degradation rates as the oxic and sulphate reduction regimes. In addition, sedimentation in the Lac du Bouchet more closely resembles the marine domain where oxic conditions prevail before degradation by sulphate reduction begins. This result indicates that oxic processes may occur at the bottom of the water column, before methanogenetic degradation, which is consistent with the seasonal oxygenation of bottom waters (Viollier *et al.*, this volume).

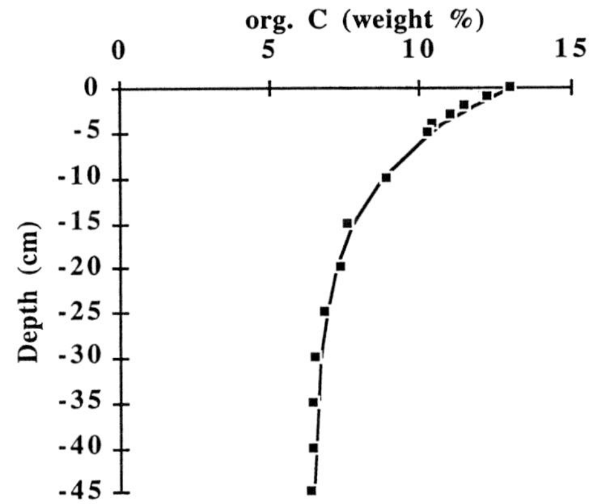

Figure 5. Model of organic carbon degradation with depth (cm) from the Lac du Bouchet. Filled squares represent measured TOC (%). The curve represents the modelled organic carbon decrease calculated with a first-order cinetic (k = 2.2 10^{-3} y^{-1}).

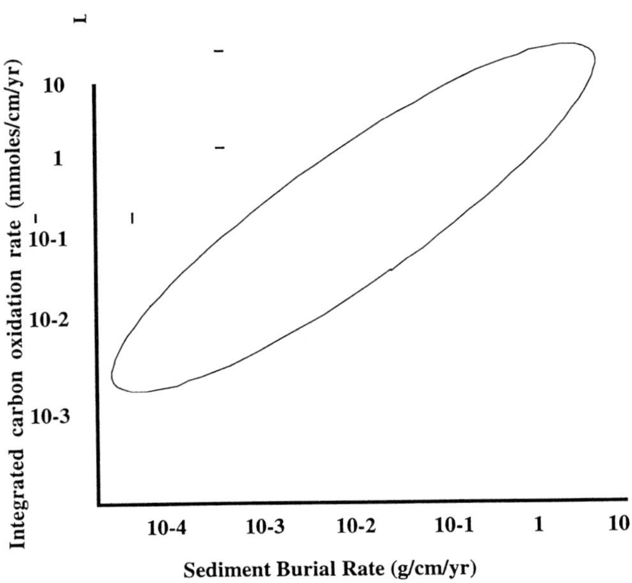

Figure 6. Diagram of the integrated carbon (40 cm) oxidation rate versus the sediment burial (sedimentation) rate for sediments from the Lac du Bouchet (solid circle) and the Lac d'Aydat (hollow circle). The envelope of oxic respiration is shown after Canfield (1989).

Organic fluxes-

The oldest of the four ^{14}C dates, 2500 ±350 years BP (at 48 cm, fig. 1a), places the sediments within the Sub-Atlantic period and illustrates the general relatively slow sedimentation rate which, however change at the boundary of each unit. In unit IA (0-1400 years BP) the total sediment flux increases slightly before increasing rapidly in unit IB (1400-1700 years BP) to three times the values in unit IA (fig.1b).

Values of total organic carbon (%TOC) were multiplied by the total sediment flux to give the organic carbon flux (mg Corg/cm^2/yr). The organic carbon flux remains relatively constant during unit IA and then increases strongly in unit IB (in parallel with the total sediment accumulation rate). It then decreases in unit II to levels close to that in unit IA (fig. 1c).

The fluxes of the alkali-soluble humic substances and their proportion in the organic material decrease during the period 0-1400 years BP (fig. 4a, 4b respectively) indicating the degradation of the autochthonous organic material during this stage. The C/N ratio in the isolated total organic material also decreases between 0-1400 years BP (fig. 4c) which highlights the relative enrichment of nitrogen in the organic material during diagenesis. The hydrogen index (HI) decreases in parallel with the C/N ratio (fig. 4c) which may be interpreted as a progressive decrease in the autochthonous/allochthonous ratio. The parallel increase of the C/N ratio and the flux of the humic substances between 1400-1700 years BP reflects the improved preservation of the organic material in these levels where the total sediment flux is greatest (due at least in part to the increased productivity).

Before 1700 years BP, the proportion of humics is relatively small whereas the C/N ratio remains high (fig. 4b and 4c). This is the result of the change in the nature of the organic matter which is much more highly degraded due to the slow accumulation rates. Proteins are probably less abundant than the resistant molecules such as lignin (ligno-cellulosic debris) and lipids (spores/pollen). These levels are characterised by relatively high HI values and low OI values which underlines the more highly hydrogenated nature of the preserved organic material (fig. 4d). The profiles of HI and OI between 0-1400 years BP reflect, therefore, the progressive degradation of the organic material during early diagenesis (fig. 4d) whilst before 1400 years BP they indicate a change in the nature of the organic material as suggested by the petrographical composition.

Conclusion-

The organic composition of the sediment during the period 0-1400 years BP records both the organic sedimentation and the effects of early diagenesis (oxic and methanogenetic processes) mainly on the autochthonous organic material. The effects of this diagenesis are only apparent because the nature of the organic input has remained relatively constant.

However, marked changes in the organic flux or in the nature of the organic matter (resulting from palaeoclimatic variations or anthropogenic influences) may mask, and indeed may retard, the effects of early diagenesis on the sediment. Therefore the impact of early diagenesis on the sedimented organic material is very variable over time because the nature and intensity of these processes are closely related to the quality (nature and state of preservation) and the quantity (flux) of each type of deposited organic material. Both these factors are extremely susceptible to even slight climatic variations.

The nature of the different petrographical organic constituents can be used as specific markers of palaeoenvironment, for example:

* opaque ligno-cellulosic debris: oxidation in the surrounding basin,
* reddish amorphous organic material: development and erosion of the soil,
* greyish amorphous organic material: primary production...

Moreover, diagenesis affects the chemical quality of the sedimentary organic material but the intensity of these effects are differential depending on the nature of the organic matter. The degree of degradation is, therefore, dominantly dependent on the sedimentation rate. Thus, the use of the organic constituents in palaeoclimatic interpretation must always integrate the differential effects of early diagenesis on each of the original components.

Acknowledgements- The authors wish to thank P. Bertrand, A. Desprairies and N. Tribovillard for their advise during this work. We would also like to thank L. Dever (Laboratoire d'Hydrogéologie et de Géochimie Isotopique, Université Paris-Sud) for radiocarbon data and D. Jalabert (Service de Microscopie de l'Université d'Orléans) for his technical assistance in TEM studies.

References-

Bertrand, P., Brocero, S., Lallier-Vergès E., Tribovillard N. and Bonifay E. (1992). Sédimentation organique lacustre et paléoclimats du Pléistocène aux moyennes latitudes: exemple du Lac du Bouchet, Haute Loire, France (preliminary results). *Bull. Soc. Géol. Fr.*, **163** (4), 427-433.

Bonifay E. and Truze E. (1987). Dynamique, sédimentation et évolution des lacs de maars: l'exemple du Velay, *In* Documents du C.E.R.L.A.T. n°1, E. BONIFAY (ed.), pp.29-64.

Bonifay E., Creer K. M. *et al.*, (1987). A study of the Holocene and Late Würmian sediments of Lac du Bouchet (Haute-Loire, France): First Results, *In* Climate - History, Periodicity and Predictability. M.R. RAMPINO CO. (ed.), New York, pp.588.

Boussafir M., Lallier-Vergès E., Bertrand P. and Badaut-Trauth D. (1994). Ultrastructural composition of selected organic matter from sediments of the Kimmeridgian Clay Formation of Yorkshire (G.B.). *Bull. Soc. Géol. Fr.*, **165**(4), 355-363.

Canfield D.E. (1989). Sulfate reduction and oxic respiration in marine sediments: implications for organic carbon preservation in euxinic environments. *Deep-Sea Research,* **36** (1), 121-138.

Derenne S., Largeau C., Berkaloff C., Rousseau B., Wilhelm C and Hatcher P.G. (1992a). Non-hydrolisable macromolecular constituents from outer walls of *Chlorella Fusca* and *Nanochlorum Eucaryotum. Phytochem.,* **31**, 1923-1929.

Derenne S., Largeau C and Hatcher P.G. (1992b). Structure of Chlorella fusca algaenans: relationships with ultralaminae in lacustrine kerogens; species- and environment-dependent variations in the composition of fossil ultralaminae. *Org. Geochem.*, **18**, 417-422.

Lallier-Vergès E., Sifeddine A., De Beaulieu, J.L., Reille M., Tribovillard N., Bertrand P., Mongenot T., Thouveny N., Disnar J.R. and Guillet B. (1993) Sensibilité de la sédimentation organique aux variations climatiques du Tardi-Würm et de l'Holocène - le lac du Bouchet (Haute Loire, France). *Bull. de la Soc. Géol. de Fr.*, **164** (5), 661-673.

Meyers P.A. and Ishiwatari R (1993) Lacustrine organic geochemistry - an overview of indicators of organic matter sources and diagenesis in lake sediments. *Org. Geochem.*, **20** (7), 867-900.

Patience A., Lallier-Vergès E., Albéric P., Tribovillard N.P. and Desprairies A. (1994) Relationships between organo-mineral supply and early diagenesis in the lacustrine Environment: a study of surficial sediments from the Lac du Bouchet (Haute Loire, France). *Quat. Sci. Rev.,* in press.

Rein B. and Negendank J.F.W. (1993). Organic Carbon contents of sediments from Lake Schalkenmehrerer Maar: a paleoclimate indicator. *In*: Lecture Notes in Earth Sciences, J.F.W. NEGENDANK and ZOLITSCHKA B. (eds.). pp. 163-170.

Sarazin, G., Michard, G., Gharib, I.Al. and Bernat, M. (1992) Sedimentation rate and early diagenesis of particulate organic nitrogen and carbon in Aydat Lake (Puy de Dôme, France), *Chem. Geol.,* **98**, 307-316.

Sifeddine A., Lallier-Vergès E., Bertrand P., de Beaulieu J.L and Reille M., (1992). Palynofaciès et flux de carbone organique au cours des 30 000 dernières années: le lac du Bouchet (Haute Loire, France). 8th International Palynological Congress, Aix en Provence, 6 - 12 sept. 1992.

Truze E. (1990). Etude sédimentologique et géochimique des dépôts du maar du Bouchet (Massif Central, France). Evolution d'un système lacustre au cours du dernier cycle climatique. Thèse de doctorat en Sciences. Université Aix-Marseille II, 242 p.

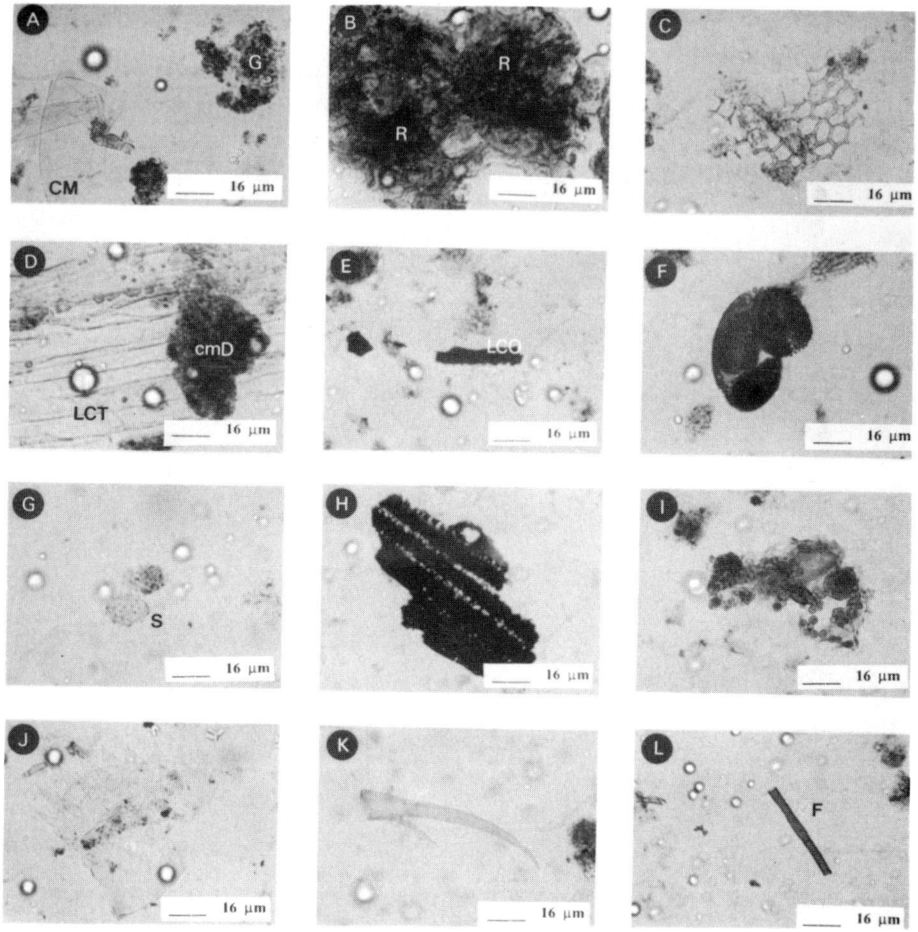

Plate 1 - Example of each type of resistant organic matter observed in palynofacies and humine preparations.
A - G: greyish amorphous organic matter (phytoplankton-derived), CM: cuticule membrane
B - R: reddish amorphous organic matter (pedogenetic) C - cuticule
D - LCT: well-preserved ligno-cellulosic debris E - LCO: oxidised ligno-cellulosic debris
F - gymnosperm pollen G - spore
H - carbonised ligno-cellulosic debris (fire-forest) I - Nitrogenous granules (pedogenetic components)
J - K - zooclasts L - Fungi filaments

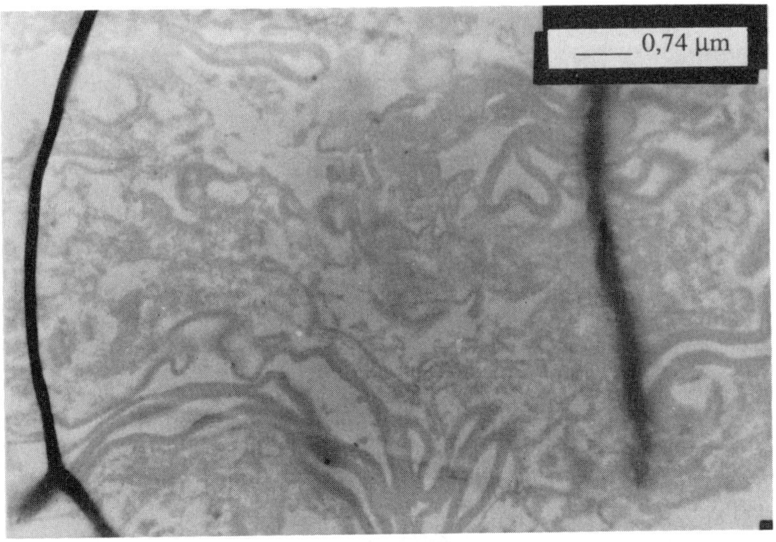

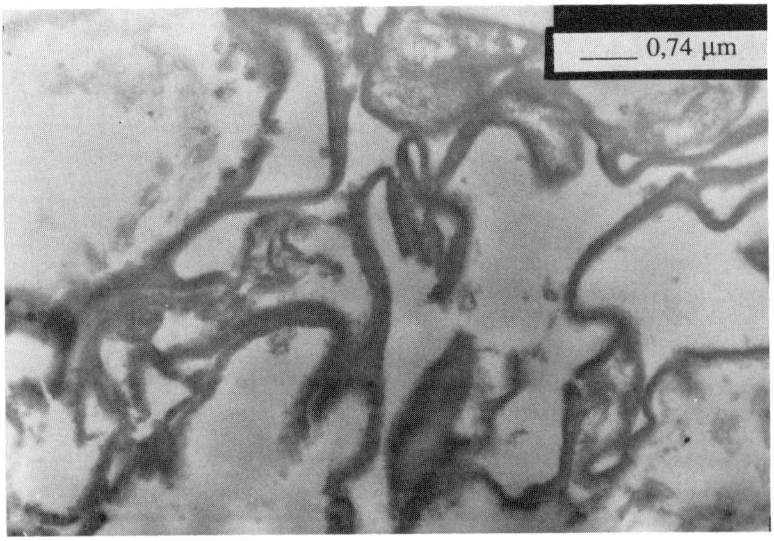

Plate 2 - Transmission electron micrographs of micro-sampled amorphous organic components. The "amorphous" organic matter, as described by light microscopy, is shown to be composed of ultralaminar structures thought to be of phytoplanktonic origin.

Organic sedimentation and its relationship with palaeoenvironmental changes over the last 30,000 years (Lac du Bouchet, Haute Loire, France). Comparison with other palaeoclimatic lacustrine examples.

Abdelfettah Sifeddine, Philippe Bertrand, Elisabeth Lallier-Vergès and Andrew Patience

Université d'Orléans, URA 724 du CNRS, Dépt. des Sciences de la Terre, F-45067 Orléans cedex

Key words- *Lac du Bouchet, organic fluxes, palynofacies, image analysis, palaeo-environment, Late Quaternary.*

Abstract- Two metres of sediments were recovered from the Lac du Bouchet (Massif Central, France) to assess the influence of climatic changes on organic lacustrine sedimentation. Palaeoenvironmental reconstruction was realised on the basis of a quantitative evaluation of the different mineral and organic fluxes. The late glacial period is marked by high mineral and low organic fluxes. The transition to the Holocene is mainly characterised by an increased flux of organic components (terrestrial biomass and phytoplankton) and a relative decrease in the mineral fluxes. The respective evolution of phytoplanktonic and terrestrial higher-plant fluxes reflects the local palaeoenvironmental variations concerning both the sedimentation and surrounding basin. The end of the Atlantic period shows a climatic cooling and the transition from Sub-Boreal to Sub-Atlantic underlines the installation of present climatic conditions. Anthropogenic influence is visible from the middle of the Sub-Atlantic. The regional palaeoenvironmental variations are also recorded in other European sites (SchalkenMehrener Maar, Lago di Monticchio), although slight differences occur due to local effects related to the geomorphology of the basins. This palaeoclimatic evolution parallels global climatic changes as recorded by the recognition of similar major trends in tropical sites (Carajas, Amazonia).

Introduction-

The Lac du Bouchet is the only maar lake in the Devès plateau which contains a continuous sedimentary record over the last 350,000 years. It is situated in the Massif Central (Haute Loire, France), in the southern part of the plateau (1200m altitude), 15 km south-west from Cayres (fig. 1). It occupies a subcircular ancient volcanic crater some 800m in diameter (Jézéquel *et al.*; Viollier *et al.*, this volume). The water depth of the lake is 28m and is controlled mainly by precipitation (meteoric waters). The surrounding catchment basin is twice the diameter of the lake and is formed from the flanks of a volcanic crater. Work already completed on the lake sediments (sedimentology, palynology, palynomagnetism, diatoms) provides a good understanding of the climatic variations over the last 120 000 years (Bonifay and Truze, 1991; Thouveny *et al.*, 1991; Reille and de Beaulieu, 1988; de Beaulieu and Reille, 1991). Bertrand *et al.* (1992) and Lallier-Vergès *et al.* (1992) showed that the organic sedimentation is very susceptible to palaeoenvironmental changes. A quantitative approach to the study of the organic inputs

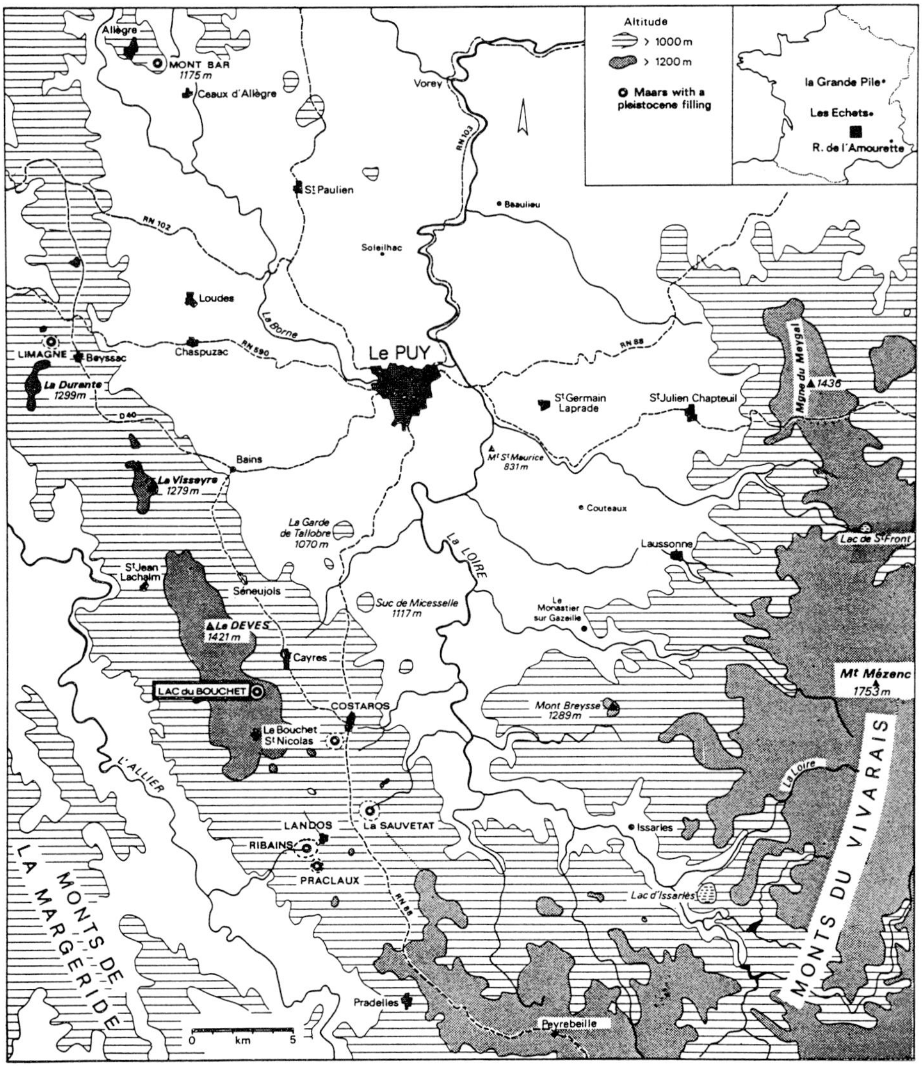

Fig. 1: Localization map of the Lac du Bouchet

has been developed using image analysis to create a framework from which it is now possible to specify how these climatic variations influence the organic sedimentation over the last 30 000 years. The comparison of the palaeoclimatic signal obtained from the Lac du Bouchet with other European and tropical examples, is attempted in order to validate the sedimentary organic matter as a palaeoenvironmental marker in the continental domain.

Samples and analytical techniques-

The present study concerns two cores recovered from the center of the Lac du Bouchet by a pneumatic corer Mackereth (LDB X) and a piston-core completed during drilling (LDB H). Sedimentological features of these cores have been already presented in Lallier-Vergès *et al.* (1993). A high resolution sampling was realised on both cores (each 2 cm for LDBH and each 5 cm for LDB X) for the geochemical and petrographical study of the organic matter.

Rock Eval Pyrolysis was performed on bulk sediment powders to obtain the total organic carbon content (%TOC) and the geochemical quality of the organic matter in terms of hydrocarbon content (hydrogen index: HI in mg HC/ g Corg).

Optical studies were performed on the total organic matter, isolated from the mineral matrix by acidic attacks (palynofacies preparation) to observe and characterise the different organic constituents and to quantity their relative proportions by image analysis (fig. 2).

A previously established stratigraphy, based on palynological and magnetic susceptibility data, allowed the estimation of average sediment fluxes (g $cm^{-2}.an^{-1}$) for the mineral and organic fractions. Due to the absence of physical parameters (water content, density) below 10m depth, the flux calculations were limited to the uppermost 10m of core LDB H which corresponding to the last 30,000 years (fig. 2 and 3).

Image analysis was performed with an OPTILAB 1.4 system coupled to a MacIntosh II. The images of palynofacies were taken by a BOSCH (CCD) high-resolution camera set on a ZEISS macroscope (with a *80 magnification). The numerisation of optical images is made with a 768*512 format and 256 grey-levels. Some artefacts, mainly due to the light heterogeneous lighting, were corrected as described in Sifeddine and Bertrand (1994). Organic particles were selected after a petrographical control and coding. The proportion of the different organic particles was obtained by considering the surfaces of each group of particle (about 300 particles have been treated). The quantitative evaluation of these organic fractions, completed on the LDB H core was compared with the visual estimation (fig. 2). The diagram shows that the two approaches give similar results concerning general trends in the variation of organic composition. However, compared to the visual estimation, the image analysis brings a better accuracy on the measurement of organic fraction proportions. Moreover, it allows, at the same time, further analyses, such as granulometry, transparency indices... (see Sifeddine and Bertrand, 1994).

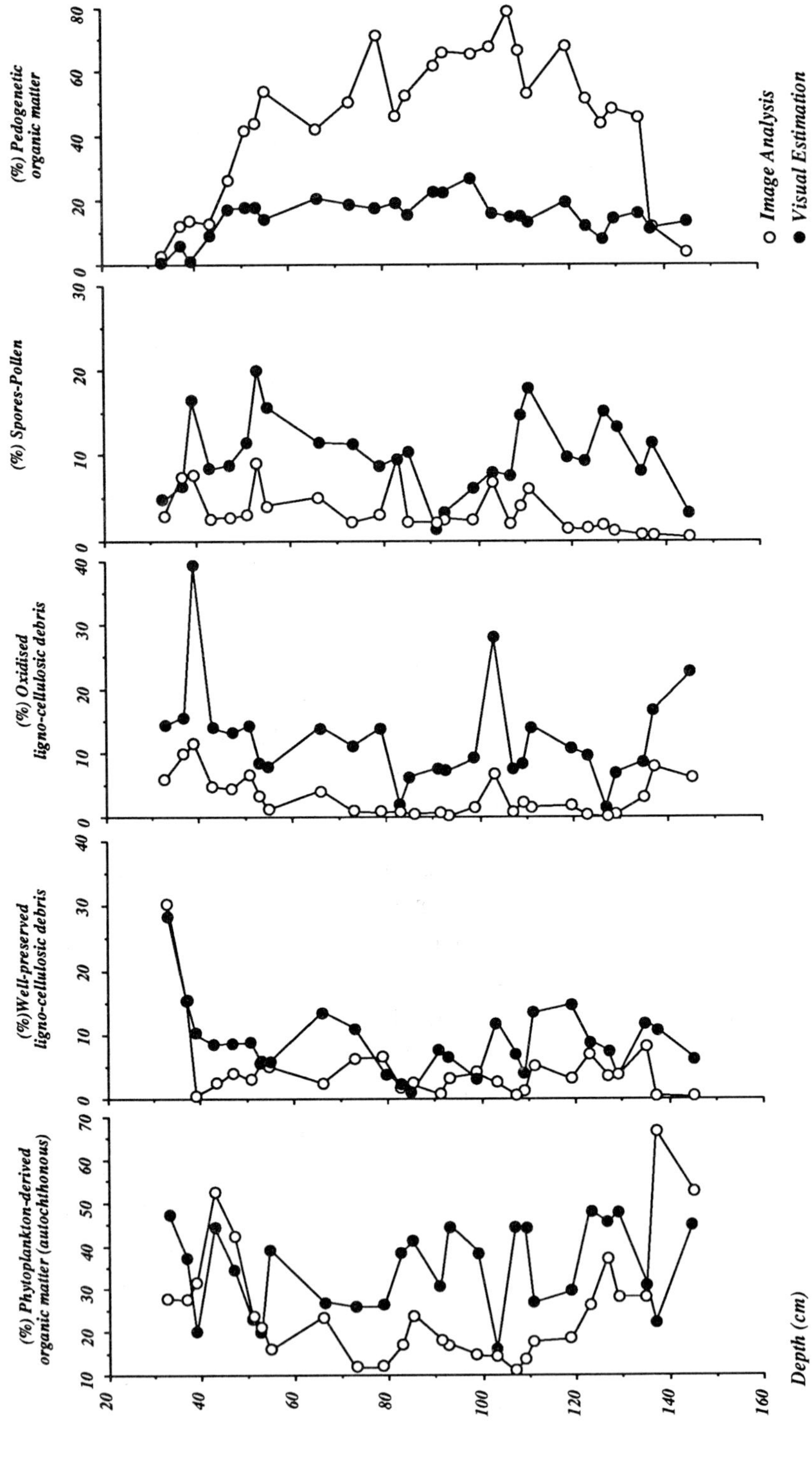

Fig. 2: Comparison between the results of palynofacies quantification by visual estimation and image analysis

Results and interpretation (fig. 3)-

From the palynofacies study, two major organic fractions, representative of the local palaeoenvironmental record, have been identified. The autochthonous organic fraction, characteristic of the sedimentation basin, is mainly composed of residual phytoplankton-derived organic matter. The allochthonous fraction, characteristic of the surrounding catchment basin, is made of an assemblage of well-preserved and oxidised ligno-cellulosic debris and pedogenetic organic material. A third organic fraction, mostly representative of the regional environment because of its mode of transportation by the winds, is composed of spores and pollen, and also by forest-fire debris. A detailed description of these organic components is presented in Patience *et al.* (this volume). The land-derived organic matter (higher plant debris) is almost always dominant throughout the core, with a high proportion of pedogenetic organic matter (fig. 2). However, two levels (at around 160cm and 50 cm down the core) are characterised by a dominance of autochthonous organic matter. The uppermost 50cm of the sedimentary column show a decrease in the pedogenetic organic fraction, relative to the ligno-cellulosic fractions, partly oxidised or well-preserved, and are essentially well-preserved in the uppermost centimetres.

The Würm period (30,000-15,000 years BP)-

During the last glacial period, as during earlier glacial ones (Bertrand *et al.*, 1992), the organic matter is not very abundant and is composed mainly of opaque ligno-cellulosic debris indicating a degraded (oxidised) material. The high mineral fluxes and the nature of the silty fraction of the sediment both suggest an intense erosion of the surrounding basin due to the weak vegetal cover. During this period, the pollen records (*Gramineae*, *Artemisia* and *Pinus*), (de Beaulieu and Reille, 1991) are characteristic of a steppic environment developed under a cool regional climate.

The Late Glacial period (15,000-9000 years BP)

In core LDB H, the Older-Dryas, Bølling-Allerød, Younger-Dryas and the Pre-Boreal are absent (Lallier-Vergès *et al.*, 1993). It is probable that this hiatus is due to the lenticular deposition which results in a non-uniform sedimentation in the lake and thus may result in irregular sediment recovery during coring in different parts of the lake. Alternatively, these episodes may simply represente very impoverished sedimentation.

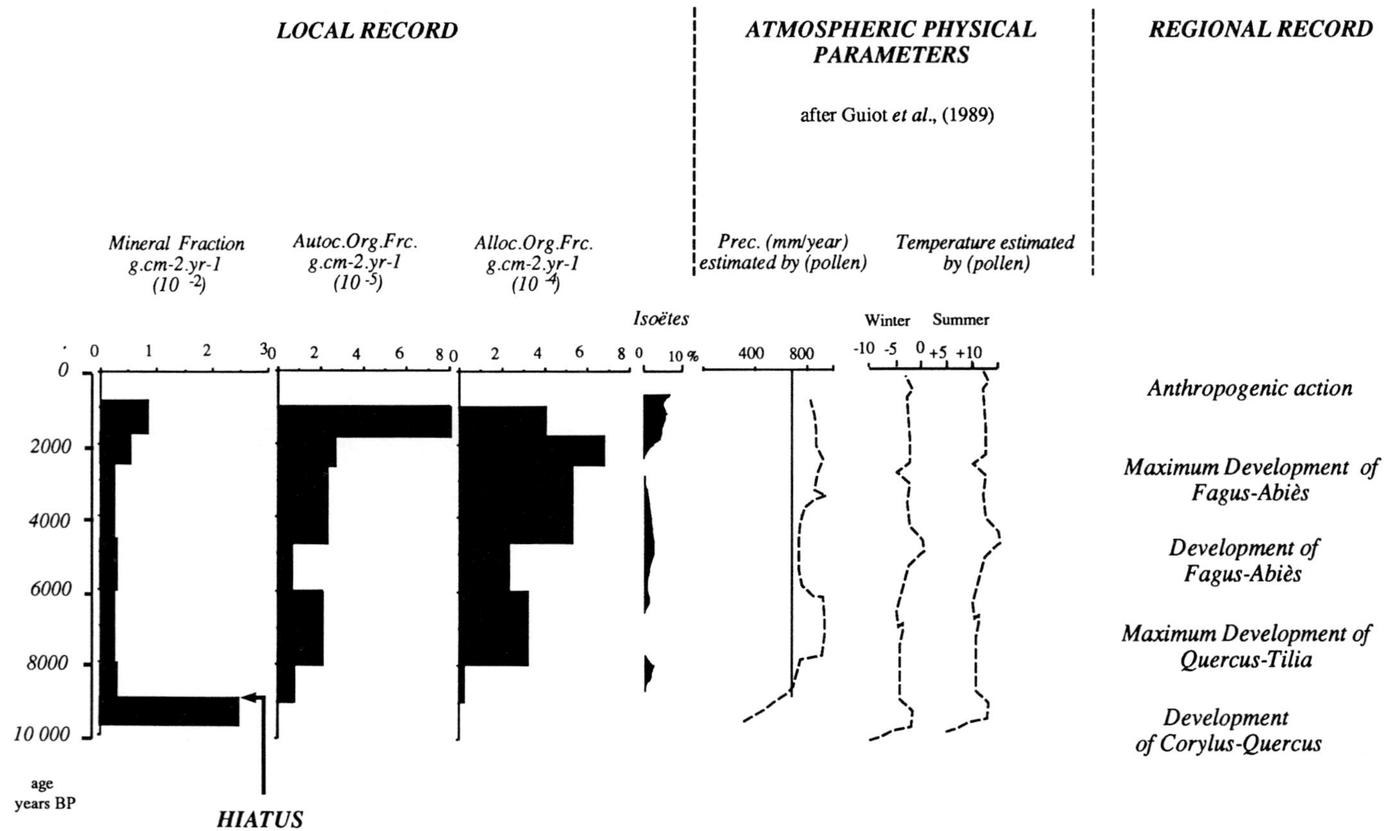

Fig. 3: Comparison between the local and regional organic records, and the physical parameters of the atmosphere at the Lac du Bouchet

The Boreal period (9000-8000 years BP)

This episode is characterised in the pollen records by the development of *Corylus* and *Quercus* forests (de Beaulieu and Reille, 1991; Reille and de Beaulieu, 1988), and is recorded in the sediment by the following:

- a decrease in the mineral fluxes,
- the improvement of the chemical quality of the organic material (increase in the HI) which can be equally well observed by the study of the palynology which shows an increase in the proportions of greyish amorphous organic material (the residual phytoplanktonic fraction), and reddish amorphous pedogenetic fraction at the expense of the opaque (oxidised) ligno-cellulosic debris,
- an increase in the total organic carbon flux, autochthonous fraction (greyish amorphous organic material) flux and allochthonous flux. The increase in the autochthonous organic fraction represented by greyish amorphous organic material (resulting from lacustrine planktonic productivity) is favoured by the input of nutrient elements during the transitional erosive phases of the surrounding basin and is interpreted as a quasi-instantaneous response of the lake to changes in the regional climatic conditions. The response of the surrounding basin (allochthonous inputs) happens only after a certain time-lag in response to the climatic amelioration (slower installation of the vegetation in the surrounding basin). The decrease in the sediment flux reflects the decrease in the erosion of the surrounding basin which is more and more protected by the installation of vegetation.

The Atlantic period (8000-4700 years BP)

The climatic optimum of the Holocene is marked in the pollen records by a forestation *(Quercus-Tilia)* period which indicates a warm and wet climate (de Beaulieu and Reille, 1991; Lallier-Vergès *et al.*, 1993). This is manifested in the organic sedimentation by moderately high total organic fluxes and by an organic material dominated by pedogenetic reddish amorphous organic matter (allochthonous). At the beginning of the Atlantic, the allochthonous and autochthonous organic fluxes increase progressively, before decreasing towards the end of this period due to a cooling of the climate (development of *Alnus -Abiès* compared to *Quercus -Tilia*). This climatic cooling had a general influence on the productivity in the lake. It directly affected the physical conditions of catchment basin and, indirectly, it caused the decrease in the input of nutrients to the lake due to a reduction in the erosion favoured by the increased vegetal cover. The progressive decrease in the percentages of aquatic plant (*Isoëtes*) pollen is interpreted as an impoverishment of the lake waters in nutritive elements, an observation which is in agreement with the weak mineral fluxes.

The Sub-Boreal period (4700-2600 years BP)

The cooling which commenced at the beginning of the Atlantic, continued into the sub-Boreal as indicated by the maximum development of the forest (*Fagus*) under a cool and humid climate. The organic sedimentation is characterised by the progressive increase in the organic carbon flux and by the increased flux of allochthonous organic material indicating the maximum development of the vegetation in the surrounding basin which stabilised the soil and minimised erosion. The low autochthonous organic fluxes may be explained by weak mineral fluxes of which the majority is inhibited by filtering through the aforested basin.

The Sub-Atlantic (2600 years BP to present)

The end of the Sub-Boreal and especially the beginning of the Sub-Atlantic (2600 years BP to recent) are characterised by the maximum organic carbon fluxes and by the abundance of the allochthonous organic fractions with high HI values. This maximum organic carbon flux is probably due to the development of a vegetation which encouraged the installation of a soil in the surrounding basin and produced, during these erosive phases, a ready supply of the pedogenetic fraction. The simultaneous increase in the mineral fraction flux and in the major element flux indicates an increased erosion of the catchment basin. This erosion is very probably caused by the installation of the recent climatic conditions approximately 2600 years BP (warming and development of less dense vegetation) which are recognised in the pollen records by the decrease in the percentages of the *Fagus* - *Quercus* pollen and an increase in the *Pinus* pollen. It is very probably the large allochthonous input that is the origin of this vegetation change. Around the middle of the Sub-Atlantic, the surrounding basin received an anthropogenic action. This was marked in the pollen records by the appearance of cereal species and by *Betula* which chacteristically pioneer the ground cleared by forest fires, recorded in the palynofacies by the presence of carbonised debris (pyrofusinite). The progressive increase in the average sedimentary fluxes prove the intensification of the erosion of the surrounding basin encouraged by the clearing of the forests. The progressive increase in the pollen of marginal aquatic plants (*Isoëtes*), and the maximum autochthonous organic flux (increased production in the lake) during the Sub-Atlantic, indicate an enrichment of nutrients in the lake waters (Shmid, 1980; Gasse, 1969).

Comparison with other palaeoclimatic records (fig. 4)-

The climatic evolution at the Lac du Bouchet (30,000 years B.P. - present) was compared to some other European palaeoclimatic examples (Schalkenmehrener Maar in Germany,

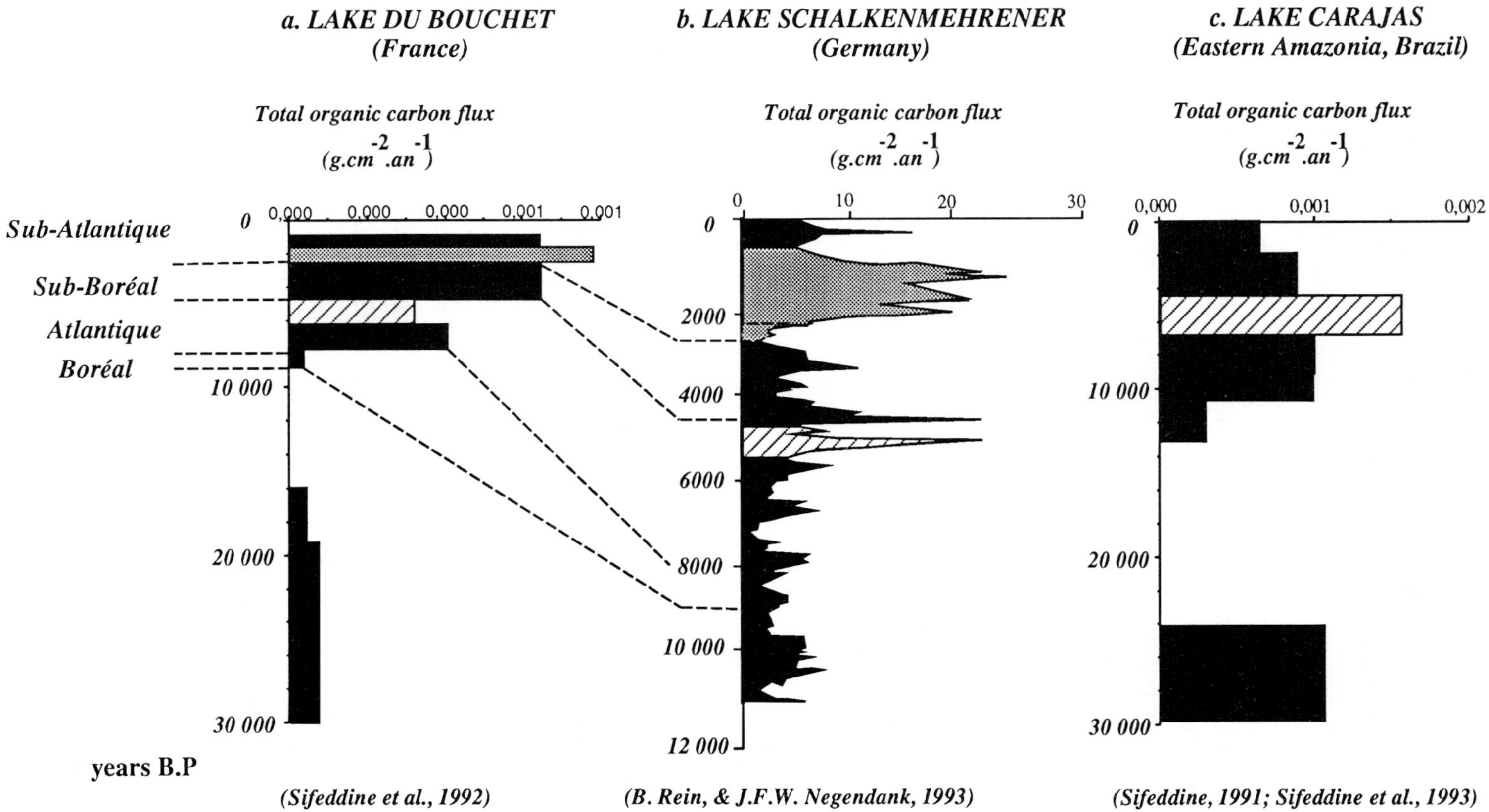

Fig. 4: Evolution of the total organic carbon flux during the last 30,000 years in the European (a,b) and South American (c) sites

Lago di Monticchio in Italy) and with an example from the tropics (Carajas in Amazonia, Brazil). Although the palaeoclimatic evolution at the Lac du Bouchet was determined using relative (palynology) and absolute (^{14}C uncorrected for $\partial^{13}C$) dates, it is clear that even in different geomorphological and climatic regimes the organic marker is directly influenced by the local conditions which are, in turn, controlled by the regional and global conditions.

The last glacial maximum at the Lac du Bouchet is characterised by a sedimentation dominantly mineral which is also observed in other regions of the world. For example, in Europe, works completed on Schalkenmehrener Maar in Germany (Rein and Negendank, 1993) and Lago di Monticchio in Italy (Zolitschka and Negendank, 1993) show that, at this time, the sedimentation was dominated by the mineral fraction. At Carajas, this episode was characterised by a sedimentary hiatus due to a complete drying out of the lake between 22,000 and 13,000 years B.P. (Sifeddine *et al.*, 1994).

Around 15,000 years BP, one notes a simultaneous recovery of the organic sedimentation in the three European sites (Lac du Bouchet, Schalkenmehrener Maar, Lago di Monticchio). At Carajas, this period is marked by an increase in the flux of total organic carbon indicating the change from an arid climate to a humide one. In south-America and Africa, this period is characterised by a general rise in the freatic nappe (Servant *et al.*, 1993). In the continental domain, this climatic amelioration (around 15,000 years B.P.) is almost synchronous with an intense global deglaciation due to the general warming of the Earth (Brocker and Denton, 1989; Lorius *et al.*, 1985).

The Holocene (10,000 - 0 years B.P.) is marked in the three European sites and at Carajas (Sifeddine, 1991) by a sedimentation which is essentially organic. Generally, the climate is humid with a climatic deterioration recorded in Europe by a short cool and dry episode around 6000 years BP, the duration of which is uncertain at the Lac du Bouchet (temperate region) and by a series of wet and dry episodes in south-America between 7000 and 4000 years B.P. (Martin *et al.*, 1992; Sifeddine *et al.*, 1993).

Acknowledgements- A special thank is devoted to the participants of the research group GdR 942 of CNRS without whose funding this work would not have been possible. The authors would like to thank also the participants of the research programs HARP, GEOCIT (ORSTOM), ECOFIT (ORSTOM/CNRS), of the research group GdR "Palaeoclimatology & Palaeohydrology" (Université de Paris Sud) and of the European program EUROMAARS for their scientific collaboration.

References-

Beaulieu J.L. de and Reille M. (1991) Analyse pollinique d'une carotte de 20 mètres du Lac du Bouchet (Velay, France). *In* "Le Lac du Bouchet: environnement naturel et étude des sédiments du dernier cycle climatique". Documents du C.E.R.L.A.T. n° 2, E. BONIFAY (ed.), pp 315-322.

Bertrand P., Brocero S., Lallier-Vergès E., Tribovillard N. and Bonifay E. (1992) Sédimentation organique lacustre et paléoclimats du Pléistocène aux moyennes latitudes: exemple du Lac du Bouchet, Haute-loire, France (résultats préliminaires). *Bull. Soc. Géol. Fr.*, **163**(4), 923-930.

Bonifay E. and Truze E. (1991) Histoire géologique du lac du Bouchet. *In* "Le Lac du Bouchet: environnement naturel et étude des sédiments du dernier cycle climatique". Documents du C.E.R.L.A.T. n° 2, E. BONIFAY (ed.), pp.35-61.

Brocker W.S. and Denton G.H. (1989) The role of ocean-atmosphere reorganizations in glacial cycles. *Geochim. Cosmochim. Acta*, **53**, 2465-2501.

Gasse F. (1969) Les sédiments à diatomées du lac Pavin (Auvergne). Annales de la Station Biologique de Besse-en-Chandesse, **4**, 221-237.

Guiot J., Pons A., de Beaulieu J.L. and Reille M. (1989) A 140 000 years continental climate reconstruction from two european pollen records. *Nature*, **338**, 309-313.

Lallier-Vergès E., Sifeddine A., de Beaulieu J.L., Reille M., Tribovillard N., Bertrand P., Montgenot T., Thouveny N., Disnar J.R. and Guillet B.(1993) Sensibilité de la sédimentation organique aux variations climatiques du Tardi-Würm et de l'Holocène - le Lac du Bouchet (Haute Loire, France). *Bull. Soc. Géol. Fr.*, **164**(5), 661-673.

Lorius C., Jouzel J. and Ritz. (1985) A 150 000 years record from Antartic ice. *Nature*, **316**, 591-596.

Martin L, Fournier M., Mourghiart P., Sifeddine A., Turcq B., Absy M.L. and Flexor J. M. (1993) Southern Oscillation Signal in South American Palaeoclimatic Data of the Last 7000 Years. *Quaternary Research*, **39**, 338-346.

Nolden G. and Nolden M. (1988) Lac du Bouchet: Results of two investigations, phytoplankton and zooplankton. Colloque du Puy en Velay, 4-6 Mai 1988.

Paillès C. (1989) Les diatomées du lac de maar du bouchet (Massif-Central, France) reconstruction des paléoenvironnement au cours des 120 derniers millénaires. Thèse d'Université. Faculté des Sciences de Luminy (Marseille, France), 272 p

Reille M. and de Beaulieu J.L. (1988). History of the Würm and Holocene vegetation in western Velay (Massif Central) : a comparison of pollen analysis from the three corings at Lac du Bouchet. *Rev. Paleobot. Palynol.*, **54**, 233-248.

Rein B.and Negendank J.F.W. (1993) Organic carbon contents of sediments from lake Schalkenmehrener maar: a paleoclimate indicator. *In* Lecture Notes in Earth Sciences, J.F.W. NEGENDANK and ZOLITSCHKA B. (eds.), Springer Verlag, Heidelberg, pp. 163-170.

Schmid A.M.M (1980) Valve morphogenesis in diatoms. *Nova Hedwigia*, **23**, 811-846.

Servant M., Maley J., Turq B., Absy M. L., Brenac P., Fournier M. and Ledru M.P. (1993) Tropical forest changes during the Late Quaternary in African and South American lowlands. *Global and Planetary Changes*, **7**, 35-47.

Sifeddine A. (1991) La sédimentation lacustre en milieu tropical (Carajas, Brésil), relation avec les changements des paléoenvironnements au cours des 60 000 dernières années. Thèse du Muséum National d'Histoire Naturelle, Paris, 119p.

Sifeddine A. and Bertrand P. (1994) Essai d'analyse quantitative des palynofaciès par traitement d'image: application sur un exemple de sédimentation lacustre - Le Lac du Bouchet (Massif Central, France). *Bull. Centres Rech. Explor. Prod. Elf-Aquitaine,* **18**, Publ. Spec., 101-106.

Sifeddine A., Bertrand P., Fournier M., Martin L., Servant M., Soubies F., Suguio K.and Turcq B. (1994) La sédimentation organique lacustre en milieu tropical humide (Carajas, Brésil): relation avec les changements des paléoenvironnements climatiques au cours des 60,000 ans B.P. *Bull. Soc. Géol. Fr.*, in press.

Thouveny N., Creer K.M and Blunk I. (1991) Extented palaeomagnetic study of the Lac du Bouchet sediments: initial reconstruction of the palaeosecularvariation over the last 100 000 years and record of post blake event palaeomagnetic excursions. *In* "Le Lac du Bouchet: environnement naturel et étude des sédiments du dernier cycle climatique". Documents du C.E.R.L.A.T. n° 2, E. BONIFAY (ed.), pp.217-248.

Truze E. (1988) Analyses chimiques des eaux du Lac du Bouchet. Communication présentée au Colloque International du Puy en Velay, 4-5-6 Mai 1988.

Zolitschka B. and Negendank J.F.W. (1993) Lago grande di Monticchio (Southern Italy): a high resolution sedimentary record of the last 70,000 years. *In* Lecture Notes in Earth Sciences, J.F.W. NEGENDANK and ZOLITSCHKA B.(eds), Springer Verlag, Heidelberg, pp. 277-288.

Sulphur speciation in the Late Glacial and Holocene sediments of the Lac du Bouchet (Haute Loire, France).

Bernard Guillet and Ousmane Maman

Université d'Orléans, URA 724 du CNRS, Dépt. des Sciences de la Terre, F-45067 Orléans cedex

Key-words- *carbon bounded sulphur, sulphate-esters, pyritic sulphur, PO_4-exchangeable sulphur, lacustrine sediment.*

Abstract- Four main sulphur forms have been identified and quantified in sediments of the Lac du Bouchet. These forms and, obviously, the total sulphur, present variations that follow those of the organic carbon. At the end of the Würm and during the Late Glacial which were periods of low productivity, sediments have low sulphur contents (<1000 µg/g) and the predominant form is the sulphate-ester fraction. During the Holocene, when the lacustrine and terrestrial biomass had developed, higher sulphur (up to 5500 µg/g) and carbon contents vary concomitantly. In Holocene sediments, the carbon bonded sulphur fraction increases to levels equal to or slightly higher than the sulphate-ester fraction. The mineral forms, identified as adsorbed sulphate and pyrite, represent a minor part of the total sulphur.

Introduction-

Sulphur dynamics have been studied for a long time in marine environments in which the microbial transformation of sulphate into sulphide leads to the formation of pyrite and to the sulphuration of organic matter. In freshwater environments and especially for lake sediments, research has been encouraged by the recent increase of the acidity of atmospheric deposition in which sulphate is a major anionic component (Cook and Schindler, 1983; Nriagu and Soon, 1985; Rudd *et al.*, 1986).

The Lac du Bouchet is a crater lake situated at the middle of a small watershed. As the lake surface represents 31% of the watershed surface (Decobert and Bonifay, 1991), the lake acts as a pluviometer (Viollier *et al.,* this volume) and collects organic and mineral particles transported over very short distances. Presently, the forest environment is composed of spruce plantings and mixed fir and beech open forests. The mean annual rainfall is 870 mm (Truze, 1990) and the concentration of sulphate in rainwater is probably close to 3 mg/l with slight interannual variations around this value. Nevertheless, it is also probable that sulphate concentrations in rainwater have varied strongly in the past due to intense volcanic activity that occurred in the closed area of the Chaîne des Puys, mainly around 8,300 yrs BP and from 6,600 to 5,700 yrs BP (Brousse *et al.*, 1969). This research deals with the forms and the distribution of sulphur in the lake sediments. Four main forms have been easily quantified on the basis of parallel and sequential extractions of sulphur fractions.

Sampling-

The Lac du Bouchet (elevation : 1205m) is a sub-circular crater lake, 28 m water depth (Viollier *et al.*; Jézéquel *et al.*, this volume) located in the Massif Central (France) at the Devès Plateau (Sifeddine *et al.*, this volume). Sediments were sampled from the LDB.X core which was recoverd from the center of the lake in September 1990. The 2.80 m long core was kept cool at 4°C before the selection of samples (5 cm thick) used for this study. The samples were freeze-dried and finely ground.

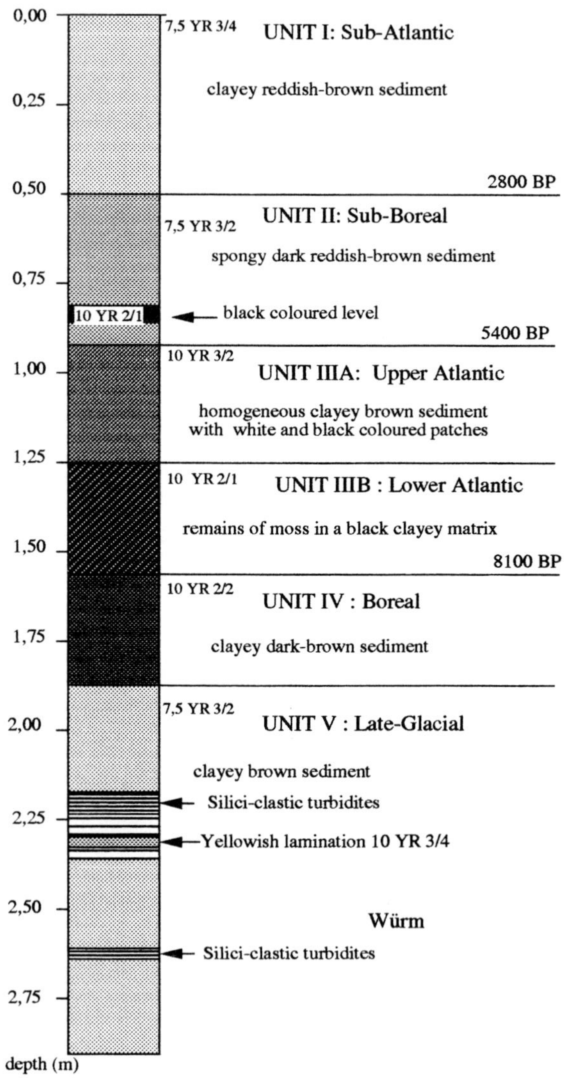

Figure 1: Description of the core LDB X. (after Lallier-Vergès et al. 1993)

Five main sedimentary units were distinguished from the color and texture of the sediment, and from analyses of palynofacies and organic matter types. The biostratigraphic chronosequence of these units was presented by Lallier-Vergès *et al.*, (1993) on the basis of the pollen analyses of the sediments. All periods of the Holocene have been recognised, except for the Preboreal. Figure 1 presents the chronosequence and the characteristics of the sediments. 18 samples were selected, namely 5, 3, 5, 2 and 3 samples for the Sub-Atlantic, Sub-Boreal, Atlantic, Boreal and Late-Glacial periods respectively.

Methods-

The chemical procedures performed for the determination and the quantification of various sulphur forms derive from the methods described by Johnson and Nishita, (1952), Zhabina and Volkov (1978) and Wieder *et al.*, (1985).

Methods of sulphur determination-

Two methods were used for sulphur determination according to the status of the material in which sulphur has to be determined. Sulphur in liquid extracts was determined in ICP atomic emission spectrometry (ICP AES) whereas sulphur in solid samples must be transformed into a volatile form (H_2S) before its determination by the method of Johnson and Nishita (J-N).

For the latter method, a sample was introduced in the boiling flask of the J-N distillation device in the presence of a reducing solution of hydriodic acid (HI) and Na-hypophosphite (NaH_2PO_2). The reaction is generated under reflux in a nitrogen flow. The released hydrogen sulphur is trapped in a solution of zinc acetate. Then, the formation of methylene blue is developed from the reaction of the zinc sulphur with a p.amino-dimethylanilin solution in the presence of $Fe^{(III)}$-NH_4 sulphate. The sulphur dosage is run by colorimetry at 670 nm.

Total sulphur and sulphur forms-

To determine total sulphur, the oxidation method using Na-hypobromide (Tabatabai, 1982) was preferred to others owing to its good repeatability.

The identified and quantified forms are presented in figure 2. It is assumed that oxidised forms such as adsorbed sulphate (SO_4^{2-}) and sulphate-esters (Est-SO_4^{2-}) may coexist with reduced forms. The latter could be mineral such as elemental sulphur (S°), metastable sulphur minerals called "acid volatile sulphur" in this paper (AVS) and pyrite (Py.S.). Organic sulphur forms in which sulphur atoms are bonded to carbon atoms (S-C) can also be present. A series of reagents used on various sediment assays allowed the evaluation of the amounts of sulphur attributable to the various forms.

a - Determination of the sulphur forms directly reducible by HI mixture.

As shown in figure 2, the reagent reduces inorganic and organic sulphate as well as elemental sulphur and dissociates labile sulphur forms (Wieder *et al.*, 1985). Samples (50 mg) were treated under reflux in the J-N distillation device for 20 minutes. Results are means of duplicate runs.

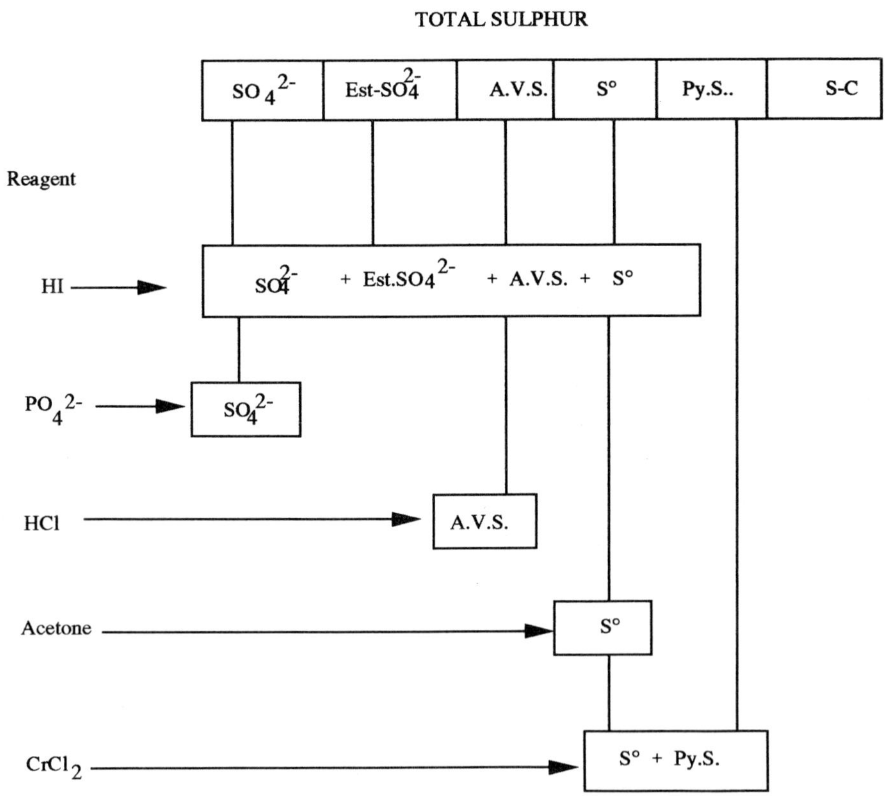

Figue 2: Flow chart of sulphur speciation analysis

b - Determination of adsorbed sulphate.

This form was extracted by a phosphate solution buffered at pH 6 (KH_2PO_4 0.1M, NaOH 0.1M and water in proportions of 50/5.6/44.4) because this reagent has been used successfully in various kinds of soils exposed to atmospheric sulphur pollution (Vannier *et al.*, 1993). The sample (500 mg) was agitated by rotation (40 r/min.) for one hour in the presence of 2.5 ml of the buffer solution. After centrifugation and filtration, sulphur of the filtrate was determined by ICP AES.

c - Determination of elemental sulphur.

The extraction procedure is similar to that described by Wieder *et al.*, (1985) using acetone as extractant. The sample (150 mg) was agitated for 16 hours with 10 ml of acetone in a hermetic flask. After filtration, 2 ml of the filtrate was introduced in the boiling flask of the J-N distillation device and the reduction of elemental sulphur was done with the HI reagent.

d - Determination of acid-volatile sulphur.

This form was determined by introducing 150 mg of sample in the boiling flask of the J-N distillation device and 15 ml of 6N HCl. The material was kept at boiling point for 30 mn. Unstable metallic sulphide compounds were decomposed and the evolved H_2S was trapped in the zinc acetate solution. After cooling, the whole suspension of the boiling flask was centrifuged. The solid residual fraction was kept as aqueous suspension in a vial under N_2 and was used for the determination of the elemental sulphur together with the pyritic sulphur.

e - Determination of pyritic sulphur.

The entire solid residue resulting from the HCl attack was immersed in the boiling flask of the N-J distillation device. Before adding 15 ml of a chromium(II) solution and 4 ml of 12N HCl, it was necessary to evacuate oxygen from the device by means of a nitrogen flow. Heat was applied after 5 minutes and the reaction time was fixed at 45 minutes. The evolved H_2S was trapped in zinc acetate solution and the dosage of sulphur was performed with the methylene blue method. The use of the chromium(II) solution has been recommended by Zhabina and Volkov (1978) and Howarth and Jorgensen (1984). This reagent was prepared from an acid solution of chromium(III) percolating through a Jones reductor following the method described in Skoog and West (1966).

Results-

Distribution of organic carbon, organic nitrogen and total sulphur-

The total organic carbon (TOC) contents vary from 6.4 to 255.0 $mg.g^{-1}$ (Table 1), with the lowest values observed in Würmian and Late-Glacial sediments (Unit V) and a peak (Unit III) corresponding to a large layer of moss stalks of *Fontinalis* (255.0 $mg.g^{-1}$). Unit I at the top of the core has TOC values of about 75 $mg.g^{-1}$.

The profile of organic nitrogen is the exact copy of TOC distribution (figure 3). In the Holocene sediment, the C/N ranges from 12 to 14 except for the level where *Fontinalis* is abundant (C/N = 19.1). An analysis of selected stalks of *Fontinalis* yields a TOC content of 358.4 $mg.g^{-1}$ and a C/N ratio of 42 which indicates that this aquatic moss is an important component of the sediment at certain levels of Unit III.

	Depth (cm)	Org. C (mg.g-1)	Org. N (mg.g-1)	Total S. (mg.g-1)	C/N atm.	N/S atm.
Sediment samples						
11211	5-15	83.40	8.10	1.87	21.0	9.9
11214	20-25	74.00	7.40	1.61	11.7	10.5
11216	30-35	76.40	7.30	1.73	12.2	9.6
11218	41-50	78.10	6.80	1.60	13.4	9.7
11220	50-55	85.50	7.40	1.77	13.5	9.6
11222	60-65	131.00	10.70	3.00	14.3	8.2
11225	75-80	162.20	n.d	3.87	n.d	n.d
11228	90-95	117.90	n.d	2.31	n.d	n.d
11230	100-105	114.00	8.50	2.03	15.7	9.6
11233	115-120	126.40	n.d	2.70	n.d	n.d
11238	140-145	255.00	15.60	4.62	19.1	7.7
11239	145-150	234.40	nd	4.70	n.d	n.d
11240	150-155	177.60	nd	5.54	n.d	n.d
11243	165-170	91.30	8.20	2.70	13.0	6.9
11247	185-190	33.30	n.d	1.30	n.d	n.d
11250	200-205	10.50	n.d	0.77	n.d	n.d
11255	225-230	7.20	1.00	0.74	8.4	3.1
11264	270-275	6.40	0.90	0.59	8.3	3.5
Fontinalis		359.10	33.70	2.99	12.4	25.7

Table 1: Organic carbon, organic nitrogen and total sulphur contents.

The total sulphur contents vary from 590 to 5535 $mg.g^{-1}$ the highest values being observed at the levels where *Fontinalis* is abundant (specially 150-155 in Unit III). A sample of *Fontinalis* collected in a mesotrophic river provided atomic C/N and N/S values of respectively 12.4 and 25.7 (TOC = 359.1 $mg.g^{-1}$). This N/S value appears to be lower than the atomic ratio value of higher plants having a modal composition in proteins and amino-acids (N/S = 34; cf. Mitchell *et al.*, 1992). As is shown in table 1, the N/S ratios of sediments are considerably lower, suggesting that sulphur forms other than proteic sulphur are present.

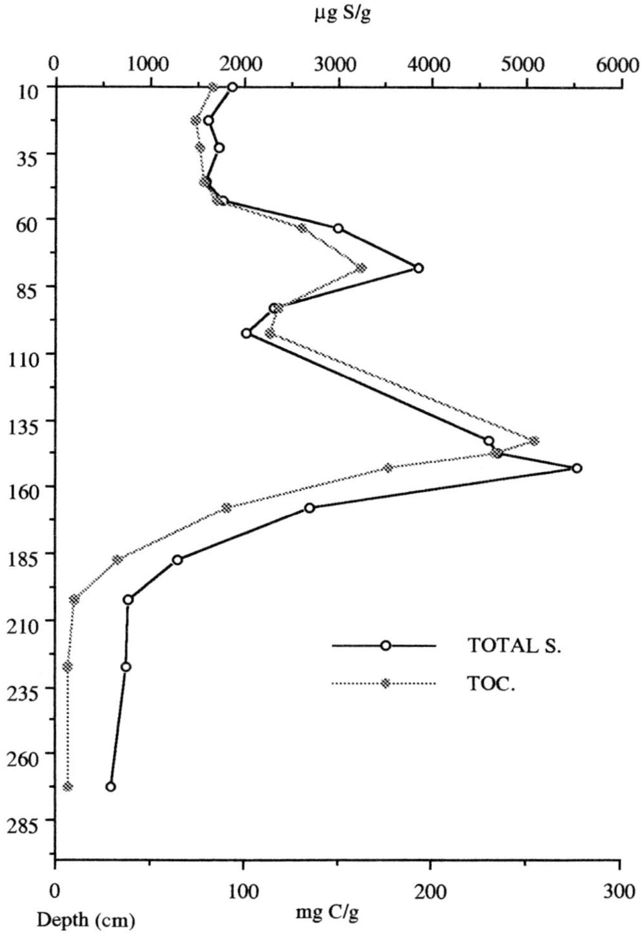

Figure 3: Variation of organic carbon and total sulphur in sediments

Inventory and quantification of sulphur forms (Table 1)-

a - PO4-exchangeable sulphate.

In the Holocene sediment, the content of PO_4-exchangeable sulphate is relatively constant with an average amount of 150 $mg.g^{-1}$ which represents nearly 8% of total sulphur. The levels from 60 to 80 cm and from 140 to 170 cm have higher values which are correlative to larger contents of total sulphur. So the maximun content is observed at the levels where the stalks of *Fontinalis* are abundant. Despite their low values in total sulphur, the Würmian and Late Glacial sediments exhibit the highest percentages of adsorbed sulphate.

Concerning *Fontinalis*, the phosphate buffer dissolved 17% of the total sulphur (510 $mg.g^{-1}$) which probably represents organic sulphur or/and sulphate salts.

b- Acid-volatile sulphur and elemental sulphur.

As the optic density of the methylene blue was very close to that of the blank, we concluded that these two sulphur forms have not been identified. In fact, traces of elemental sulphur (S°) and acid-volatile sulphur (AVS) might be present at the top and at the bottom of the core respectively.

c-Sulphur forms reducible by the HI-Phosphite reagent : evalution of the sulphate-ester fraction.

The profile of this sulphur compartment has the same design as that of total sulphur. This compartment includes PO4-exchangeable sulphate (SO_4^{2-}), sulphate-esters (Est-SO_4^{2-}), elemental sulphur (S°) and acid-volatile sulphur (AVS).

$$(\text{S-HI}) = (\text{Est-}SO_4^{2-}) + (SO_4^{2-}) + (S°) + (\text{AVS}).$$

As the elemental sulphur (S°) and the acid-volatile sulphur (AVS) have not been identified, the sulphate-ester fraction was deduced from the following equation :

$$(\text{Est-}SO_4^{2-}) = (\text{S-HI}) - (SO_4^{2-})$$

Figure 4 shows that sulphate-esters are a sizeable sulphur fraction, varying from 28 % to 62 % of the total sulphur. The percentage of this fraction is much higher in the Late Glacial and Würmian sediments than in the Holocene sediments.

The aquatic plant *Fontinalis* contained 490 ppm of reducible sulphur by HI. The similarity of results obtained by HI analysis and in the extract of the phosphate buffer (510 ppm) suggests that there is only one compartment of the oxidised sulphur, containing probably sulphated salts. These represented 17% of total sulphur.

d - Pyritic sulphur.

The mean content was about 50 μg.g^{-1} except in some levels rich in organic matter where the values can reach several hundred μg per gram of sediment (table 2 and figure 4). It is well known that the organic matter is an essential constituent in the development of the sulphate reduction and consequently to the genesis of pyrite. However there is no exact relationship between the organic carbon content and the pyrite content of sediment. For example, at the 100-105 cm level, the organic carbon content (114 μg/g) is slightly lower than at the 60-65 cm level (131 μg/g) whereas the pyritic sulphur content is considerably lower (35 μg.g^{-1} instead of 178 μg.g^{-1}). This indicates that the genesis of pyrite is considerably influenced by the nature of the organic matter.

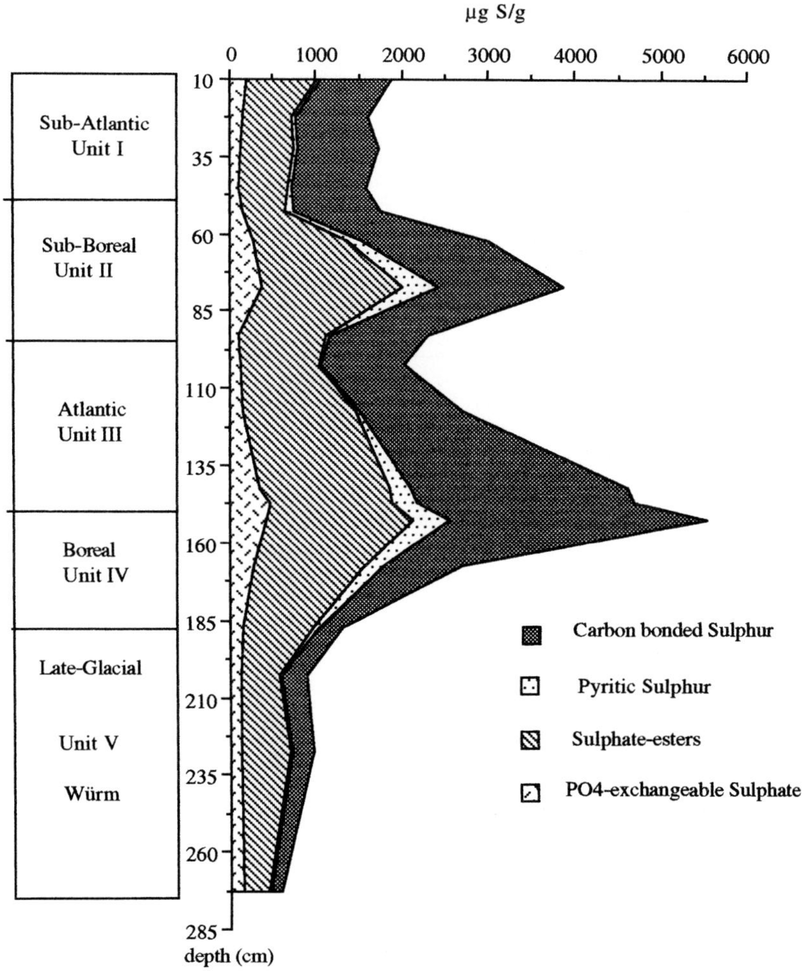

Figure 4: Variation of the Sulphur forms

Sediment samples	Depth (cm)	Total S. (µg.g-1)	SO4 (µg.g-1)	SO4 (%)	S - HI (µg.g-1)	S - HI (%)	Est. SO4 (µg.g-1)	Est. SO4 (%)	Py. S. (µg.g-1)	Py. S. (%)	S- C (µg.g-1)	S- C (%)
11211	5-15	1875	195	10	1010	54	815	44	50	3	815	43
11214	20-25	1610	160	10	730	45	570	35	40	2	840	52
11216	30-35	1735	115	7	755	44	640	37	35	2	945	54
11218	41-50	1595	105	7	675	42	570	35	45	3	875	55
11220	50-55	1770	160	9	650	37	490	28	90	5	1030	58
11222	60-65	2995	275	9	1350	45	1080	36	180	6	1465	49
11225	75-80	3870	365	9	2020	52	1655	43	410	11	1440	37
11228	90-95	2310	110	5	1125	49	1015	44	40	2	1145	50
11230	100-105	2030	115	6	1030	51	915	45	35	2	965	48
11233	115-120	2695	150	6	1455	54	1305	48	45	2	1195	44
11238	140-145	4620	355	8	1870	40	1515	32	230	5	2520	55
11239	145-150	4695	470	10	1895	40	1425	30	275	6	2525	54
11240	150-155	5535	420	8	2135	39	1715	31	420	9	2980	54
11243	165-170	2700	280	10	1560	58	1280	48	215	8	925	34
11247	185-190	1305	160	12	975	75	815	63	70	6	260	20
11250	200-205	885	130	15	565	64	435	49	30	4	290	33
11255	225-230	970	125	13	705	73	580	60	25	2	240	25
11264	270-275	590	140	23	435	74	295	51	25	4	130	22
Fontinalis		2995	510	17	490	16			22	0.7	2505	84

Table 2: The various sulphur forms: contents and percentages of total Sulphur.

In *Fontinalis* samples, traces of sulphur have been identified as pyrite (0.7% of the total sulphur), but this slight amount probably derives from the breakdown of organic sulphur and from the distillation of volatile sulphur.

e - Carbon bonded sulphur.

This form results from the difference between the total sulphur and the sum of HI-reducible sulphur and pyritic sulphur. It represents sulphur amounts that are almost equivalent to those reducible by the HI mixture. However, in Würmian and Late Glacial sediments, the S-C forms are slightly less abundant than HI-reducible forms.

Using the N content of *Fontinalis* presented in table 1 and the value of the S-C fraction (2.50 mg/g), one finds a N/S atomic ratio of 31 that is more compatible with the nitrogen- and sulphur-bearing amino acids of plants.

Discussion-

The sulphur variations replicate the organic carbon variations (fig. 3). This parallelism is the result of the productivity and biomass changes that have occurred in the lake itself and in its watershed since the end of the Würm. Whereas the Würmian and Late Glacial sediments have low contents in organic carbon and sulphur, the Holocene sedimentation clearly shows two peaks for each element. The first peak (Unit III) corresponds to the base of Atlantic period which was warmer and more rainy than the present (Reille and de Beaulieu, 1988). The development of a dense mixed-oak forest in the watershed probably led to an important leaching of nutrients from the soils. The consequence was an increase of the primary productivity of aquatic organisms. The high content in organic matter and sulphur, especially in organic sulphur, appears to be related to the abundant presence of plant remains such as the aquatic moss *Fontinalis*. The growth of this moss in the euphotic zone requires support such as immersed rocks or tree trunks, which is unrealistic considering the water-depth at the center of the lake. Thus, its sedimentation is thought to be due to transport mechanisms. The second peak was found in sediments dated of the Sub-Boreal period (Unit II) in which beech forest replaced the oak forest

The organic sulphur forms, including sulphate-esters and carbon bonded sulphur represent the largest part of the total sulphur. The mineral forms identified as PO_4^{2-} exchangeable adsorbed sulphate and pyritic sulphur are minor fractions. This sulphur distribution confirms previous results obtained in various lake sediments (Nriagu and Soon, 1985). However, slight differences appear in the relative distribution of organic and mineral sulphur according to the environmental conditions of sedimentation.

In the Würmian and Late Glacial sediments, the sulphate-esters constitute the most important compartment of organic sulphur and predominate over the carbon bonded sulphur compartment. As sulphate-esters are known to be microbially mediated in soil environment (Fitzgerald, 1976), one can assume that the sulphate-esters are also biosynthesised into lacustrine sediments. The low C/N (8, cf. table 1) of the organic matter is in agreement with the hypothesis of a large proportion of organic matter of microbial origin.

During the Holocene and especially after the Boreal period, the dominating carbon bonded sulphur forms appear in relation with inherited organic matter. The parallel increase in organic carbon and sulphur during the Holocene is to be related to the development of biomass in the watershed as well as in the lake itself and to the transport of partially degraded remains and humified compounds deriving from the soils as assessed by the optical study of the sedimentary organic matter (Lallier-Vergès *et al.*; 1993; Patience *et al.*, this volume). In such humified products, sulphate-esters can be present especially in fulviç acids (Vannier and Guillet, 1994) and linkages between sulphur and carbon atoms may exist under more stable forms than those existing in amino acids. Knowing the low productivity of terrestrial ecosystems in periglacial environments (Gore, 1983), we can assume according to Sifeddine *et al.* (this volume) that during the Late Glacial period, very few humified remains of higher plants were transported towards the bottom of the lake.

Sulphur in oxidised state, as adsorbed sulphate and sulphate-esters, represents 65 to 75% of the total sulphur of Würmian and Late Glacial sediments and 35 to 60 % of total S in Holocene sediments. Given that careful precautions were taken during the sampling and that the oxidation of the material was negligible, these high values of oxidised sulphur preclude the hypothesis of strongly anoxic conditions in the sediments.

However, pyritic sulphur is present in sediments where a large proportion of carbon bonded sulphur forms was found (fig. 4). Pyrite is mainly located inside cells of vegetal remains as has often been observed in sediments of the Lac du Bouchet (Patience *et al.* , this volume). In such microsites, the formation of pyrite was efficient because of the development of local microbial sulphate reduction.

Conclusion-

The organic sulphur forms are considerably more abundant than mineral sulphur forms corresponding to adsorbed sulphate and pyrite. In the Würmian and Late Glacial sediments, sulphate-esters are the main organic sulphur forms, and their formation appears to be related to the presence of organic matter, probably of microbial and phytoplanktonic origin rather than terrestrial origin. During the Holocene, the influence of the aquatic ecosystems of the lake shore and of the forest ecosystems contributes to the concomitant increase of carbon and sulphur in sediments. The organic forms in which sulphur is linked to carbon notably increase. From a geochemical point of view, the sedimentation conditions do not seem to have been highly anoxic, since the adsorbed sulphate and the sulphate-esters represent almost half of total sulphur in Holocene sediments and even much more in Würmian and Late Glacial sediments.

Acknowledgements- Thanks are due to Françoise Champion for her help during the sampling and the preparation of the samples.

References-

Brousse R., Delibrias G., Labeyrie J. and Rudel A. (1969). Eléments de chronologie des éruptions de la Chaine des Puys. *Bull. Soc. Géol. France*, **11**(7), 770-793.

Cook R.B. and Schindler D.W. (1983). The biogeochemistry of sulfur in an experimentally acidified lake. *Ecol. Bull.*, **35**, 115-127.

Decobert M.and Bonifay E. (1991). Bathymétrie et géomorphologie du lac du Bouchet et de son bassin versant. *In*: " Le Lac du Bouchet (I) : environnement naturel et étude des sédiments du dernier cycle climatique" E. BONIFAY (éd.), Document du C.E.R.L.A.T., n°2, pp.79-88.

Fitzgerald J.W. (1976). Sulfate ester formation and hydrolysis : a potentially important yet often ignored aspect of the sulfur cycle of aerobic soils. *Bacterial. Rev.*, **40**, 698-721.

Gore A.J.P. ed., (1983). Mires : swamp, bog fen and moor. General studies. *Ecosystems of the World,* vol. 4A, Elsevier publish., 440p.

Howarth R.W. and Jorgensen B.B. (1984). Formation of ^{35}S-labelled elemental sulfur and pyrite in coastal marine sediments (Limfjorden and Kysing Fjord, Denmark) during short-term $^{35}SO_4^{2-}$ reduction measurements. *Geochim. Cosmochim. Acta,* **48**, 1807-1818.

Johnson C. M. and Nishita H. (1952). Microestimation of sulphur in plant material, soils and irrigation water. *Analytical Chemistry*, **24**, 736-742.

Lallier-Vergès E., Sifeddine A., de Beaulieu J.L., Reille M., Tribovillard N., Bertrand P., Mongenot T., Thouveny N., Disnar J.R. and Guillet B. (1993) Sensibilité de la sédimentation organique aux variations climatiques du Tardi-Würm et de l'Holocène; le Lac du Bouchet (Haute-Loire, France). *Bull. Soc. Géol. Fr.*,**164**(5), 661-673.

Mitchell M.J., Harrison R.B., Fitzgerald J.W., Johnson D.W., Lindberg S.E, Zhang Y. and Autry A. (1992) Sulfur chemistry, deposition and cycling in forest : Sulfur distribution and cycling in forest ecosystems. *In*: "Atmospheric deposition and forest nutrient cycling" D.W. JOHNSON and S.E. LINDBERG (eds.), Springer-Verlag Inc. Publish., pp. 90-128.

Nriagu J.O. and Soon Y.K. (1985) Distribution and isotopic composition of sulfur in lake sediments of northern Ontario. *Geochim. Cosmochim. Acta*, **49**, 823-834.

Reille M. and de Beaulieu J.L. (1988) History of the Würm and Holocene vegetation in western Velay (Massif Central): a comparison of pollen analysis from the three corings at Lac du Bouchet. *Rev. Paleobot. Palynol.*, **54**, 233-248.

Rudd J.W., Kelly C.A. and Furutani A. (1984) The role of sulfate reduction in long term accumulation of organic and inorganic sulfur in lake sediments. *Limnol. Oceanogr.*, **31** (6), 1281-1291.

Skoog D.A. and West D.M. (1966) Fundamentals of analytical chemistry. Holt, Reinhart and Winston, 786p.

Tabatabai M.A. (1982) Sulfur. *In* "Methods of soil analysis, Part 2: chemical and microbiological properties", Am. Soc. Agron. Madison Wisco., Agron Serie 9, 2° ed., Ch 28, 501-538.

Truze E. (1990) Etude sédimentologique et géochimique des dépôts du maar du Bouchet (Massif Central, France). Evolution d'un système lacustre au cours du dernier cycle climatique (0-120 000 ans). Thèse de l'Université d'Aix-Marseille III, 242 p.

Vannier C., Didon-Lescot J.F., Lelong F. and Guillet B. (1993) Distribution of sulphur forms in soils from beech and spruce forests of Mont-Lozère (France). *Plant and Soil*, **154**, 197-209.

Vannier C. and Guillet B. (1994) Sulphur forms in the organic fractions of an upland forest soil (Mont-Lozère-France). *Soil Biol. Biochem.*, **26**(1) 149-151.

Wieder R.K., Lang G.E. and Granus V.A. (1985) An evaluation of wet chemical methods for quantifying sulfur in freshwater wetland peat. *Limnol. Oceanogr.*, **30**(5), 1109-1115.

Zhabina N.N. and Volkov I.I. (1978) A method of determination of various sulfur compounds in sea sediments and rocks. *In* "Environmental biogeochemistry and geomicrobiology". Vol. 3: Methods, Metals and Assessment, W. E. KRUMBEIN (ed.), Ann Arbor Science publish., pp. 735-746.

List of contributors-

Patrick Albéric, Université d'Orléans, URA 724 du CNRS, Dépt. des Sciences de la Terre, F-45067 Orléans cedex

El Mostafa Bachaoui, Université Paris Sud, URA 723 du CNRS, bâtiment 504, F-91405 Orsay cedex

Denise Badaut-Trauth, Muséum National d'Histoire naturelle de Paris, URA 723 du CNRS, 43, rue Buffon, F-75005 Paris

Philippe Bertrand, Université d'Orléans, URA 724 du CNRS, Dépt. des Sciences de la Terre, F-45067 Orléans cedex
present address: Université de Bordeaux, URA 197 du CNRS, Dépt. de Géologie et d'Océanographie, F-33405 Talence cedex

Mohammed Boussafir, Université d'Orléans, URA 724 du CNRS, bâtiment Géosciences, F-45067 Orléans cedex

Sylvie Derenne, Ecole Nationale Supérieure de Chimie de Paris, 11 rue Pierre et Marie Curie, F-75231 Paris cedex 5

Alain Desprairies, Université Paris Sud, URA 723 du CNRS, bâtiment 504, F-91405 Orsay cedex

Jean-Robert Disnar, Université d'Orléans, URA 724 du CNRS, bâtiment Géosciences, F-45067 Orléans cedex

Marc Evrard, Laboratoire de Géochimie des Eaux, Université Paris VII, case postale 7052, F-75251 Paris cedex 5

François Gelin, Ecole Nationale Supérieure de Chimie de Paris, 11 rue Pierre & Marie Curie, F-75231 Paris cedex 5

Bernard Guillet, Université d'Orléans, URA 724 du CNRS, Dépt. des Sciences de la Terre, F-45067 Orléans cedex

Didier Jézéquel, Laboratoire de Géochimie des Eaux, Université Paris VII, case postale 7052, F-75251 Paris cedex 5

Elisabeth Lallier-Vergès, Université d'Orléans, URA 724 du CNRS, Dépt. des Sciences de la Terre, F-45067 Orléans cedex

Claude Largeau, Ecole Nationale Supérieure de Chimie de Paris, 11 rue Pierre et Marie Curie, F-75231 Paris cedex 5

Dominique Lavergne, Laboratoire de Géochimie des Eaux, Université Paris VII, case postale 7052, F-75251 Paris cedex 5

Ousmane Maman, Université d'Orléans, URA 724 du CNRS, Dépt. des Sciences de la Terre, F-45067 Orléans cedex

Gil Michard, Laboratoire de Géochimie des Eaux, Université Paris VII, case postale 7052, F-75251 Paris cedex 5

Andrew Patience, Université d'Orléans, URA 724 du CNRS, Dépt. des Sciences de la Terre, F-45067 Orléans cedex

Monique Pèpe, Laboratoire de Géochimie des Eaux, Université Paris VII, case postale 7052, F-75251 Paris cedex 5

Lalanirina Ramanampisoa, Université d'Orléans, URA 724 du CNRS, Dépt. des Sciences de la Terre, F-45067 Orléans cedex

Abdelkader Ramdani, Université Paris Sud, URA 723 du CNRS, bâtiment 504, F-91405 Orsay cedex

Gérard Sarazin, Laboratoire de Géochimie des Eaux, Université Paris VII, case postale 7052, F-75251 Paris cedex 5

Abdelfettah Sifeddine, Université d'Orléans, URA 724 du CNRS, Dépt. des Sciences de la Terre, F-45067 Orléans cedex
present address: ORSTOM, UR 1C, 72 route d'Aulnay, 93143 Aulnay cedex

Nicolas-Pierre Tribovillard, université Paris Sud, URA 723 du CNRS, bâtiment 504, F-91405 Orsay cedex

Eric Viollier, Laboratoire de Géochimie des Eaux, Université Paris VII, case postale 7052, F-75251 Paris cedex 5

Pierpaolo Zuddas, Laboratoire de Géochimie des Eaux, Université Paris VII, case postale 7052, F-75251 Paris cedex 5

Subject Index

accumulation of organic matter 7, 21, 49, 58, 78
alginite 20, 30
alkalinity 104, 125, 127
aluminium 129, 134
ammonium 100, 102, 125, 128, 131, 132, 134
amorphous organic matter 4, 5, 9, 16, 17, 18, 20, 21, 31, 43, 44, 148, 149
arsenic 107
Atlantic 163, 179
bacteranes 44
barium 106, 116
beidellite 77
bio-opal 124, 133
biogeochemical processes 95, 110, 114
bituminite 20, 30
Bolling-Allerod161
Boreal 163, 180
Boulonnais 82
burial depth 78
burial temperature 78
calcium 135
Carajas 165
carbonate precipitation 79
catchment basin 148
clay chemistry 70, 71, 79, 84
clay mineral assemblage 19, 50, 63, 122
clay mineral diagenesis 77, 78, 85
clay morphology 70
clay structural formulae 71, 77
Cleveland Basin 3, 11, 19, 22, 31, 49, 63
cobalt 106
composition of organic matter 4
copper 117
datation (^{14}C) 152
degradation of organic matter 5, 7, 9, 10, 21, 40, 55, 58, 59, 60, 79, 95, 114, 124, 131, 132, 133, 136, 149, 150
degree of pyritisation (DOP) 9, 80
depositional conditions 4, 9, 15, 49, 50, 60, 78
diagenesis 77, 78, 85, 119, 145, 153
diatoms 149
dinoflagellates 55
dissolved organic matter 125, 126
Dorset 7, 11, 19, 82
epilimnion 97
Gas chromatography 51
GC-MS 32, 36, 39, 51
Haute-Loire, France 95, 119, 145, 157, 169
Holocene 95, 119, 145, 157, 169
humine 148
hydrocarbons 5, 7, 9, 10, 11, 52, 58
hydrogen sulphide 6, 9, 10, 21
hypolimnion 100, 114
illite 66-85
inertinite 53
infrared spectroscopy 66, 79
inorganic dissolved carbon 112, 113

interstitial water 95, 114, 121, 136, 138-143, 149
iron 11, 18, 21, 70, 78, 79, 100, 101, 114, 127, 134
kaolinite 66-85
Kimmeridge Clay Formation 3, 11, 19, 22, 31, 49, 63
Lac du Bouchet 95, 119, 145, 157, 169
lac d'Aydat 150
Lago di Monticchio 165
lake drainage 96
lake morphology 96
land-derived organic matter 5, 17, 43
Late glacial 161, 174, 179, 180, 181
lipidic organic matter 36, 40, 45, 58, 59, 60
maar lake 95, 119, 145, 157, 169
magnesium 135
manganese 100, 102, 128
mass-balance calculation 79
metabolisable organic matter 6, 7, 9, 11, 22, 44, 58, 78
methanogenesis 125, 133, 149
methylsterane index 55, 56
mica 66-85
microbial degradation 5, 7, 9, 10, 21, 40, 55, 58, 59, 60, 79
microfacies 19
mineralization of organic matter 131-133
mineralogy 66-85
mixed-layer illite/smectite 66-85
modelisation 9, 10, 50
molecular biomarkers 21, 33-43, 49-60
neoformed kaolinite 78, 79, 80
Nitrogen 174, 175, 180
nutrients 100
offline pyrolysis 32-34
Older Dryas 161
organic carbon cyclicity 3, 4, 5, 6, 9, 22, 31, 44, 45, 50, 51, 58, 63
organic fluxes 9, 11, 22, 55, 145, 152, 153
organic matter maturity 52
organic matter nature 148
organic sulphur compounds 5, 10, 43
oxide reduction 124
oxygen content 99, 100, 114
palaeoclimate 11, 58, 80, 82, 146, 148, 153, 157, 161, 163, 164, 166
palynofacies observation 16, 20, 148, 149, 160, 161
pedogenesis 148, 149, 160
pH 96, 101, 116, 127
phosphorus 100, 125, 129, 132
phytoplankton 5, 9, 10, 22, 49, 58, 60, 156, 160, 161
potassium 135
Pre-Boreal 161
preservation of organic matter 7, 15, 22, 52, 55, 78, 161
primary production 7, 9, 11, 15, 44, 45, 53, 55, 58, 60, 95, 114
pyrite 5, 7, 18, 19, 20, 22, 30, 59, 124, 134, 149, 169, 177, 180
pyrolysis-GC-MS 32, 36
quartz 124
rainfall 104, 105, 146
redox conditions 9, 78, 95, 114
resistant organic matter 5, 22, 36, 45, 55, 58, 155
Schalkenmehrener Maar 165
sea level fluctuation 64, 82, 85
sediment porosity 122

sedimentation rate 63, 78, 95, 119, 122, 136, 150
selective preservation 5, 7, 40, 43, 44
SEM 15, 66
sequential-leach extraction 84
short-term climatic oscillation 82
siderite 134
silica 100, 103, 129, 133
source rocks 3
STEM 18
strontium 104
Sub-Atlantic 152, 164
Sub-Boreal 164, 179
sulphate 4, 5, 6, 7, 9, 11, 18, 19, 21, 22, 44, 45, 55, 58, 78, 103, 104, 124, 128, 133, 149, 169-176, 180
sulphate reduction 4, 5, 6, 7, 9, 11, 18, 19, 21, 22, 44, 45, 55, 58, 78, 103, 104, 124, 128, 133, 149, 169-176, 180
sulphate reduction index (SRI) 6, 7, 21
sulphide 4, 5, 6, 7, 9, 11, 18, 19, 21, 22, 44, 45, 55, 58, 78
sulphur 4, 5, 6, 7, 9, 11, 18, 19, 21, 22, 44, 45, 55, 58, 78, 149, 150, 151, 169, 171-180
superficial sediment 119, 145
Tasmanacae 20, 30
TEM 5, 15, 17, 20, 30, 31, 43, 44, 66, 70, 156
temperature 96, 99
third order cycles 63, 80, 82
thorium 84
trace elements 4, 85, 95, 104, 114
ultralaminae 5, 17, 31, 39, 43, 45, 156
ultrastructural feature of organic matter 17 - 21, 30
UV excitation 20, 30
vanadium 107, 117
vivianite 134
vulcanisation 5, 10, 11, 36, 40, 43, 44, 45
water balance 95, 108, 109, 136
water chemical properties 98
water column stability 146
water physical properties 98
water stratification 97
Wurm 161, 174, 179, 180, 181
x-ray diffraction 66, 70, 71, 79
Yorkshire 3, 11, 19, 22, 31, 49, 63
Younger Dryas 161

Springer-Verlag and the Environment

We at Springer-Verlag firmly believe that an international science publisher has a special obligation to the environment, and our corporate policies consistently reflect this conviction.

We also expect our business partners – paper mills, printers, packaging manufacturers, etc. – to commit themselves to using environmentally friendly materials and production processes.

The paper in this book is made from low- or no-chlorine pulp and is acid free, in conformance with international standards for paper permanency.

Lecture Notes in Earth Sciences

Vol. 1: Sedimentary and Evolutionary Cycles. Edited by U. Bayer and A. Seilacher. VI, 465 pages. 1985. (out of print).

Vol. 2: U. Bayer, Pattern Recognition Problems in Geology and Paleontology. VII, 229 pages. 1985. (out of print).

Vol. 3: Th. Aigner, Storm Depositional Systems. VIII, 174 pages. 1985.

Vol. 4: Aspects of Fluvial Sedimentation in the Lower Triassic Buntsandstein of Europe. Edited by D. Mader. VIII, 626 pages. 1985. (out of print).

Vol. 5: Paleogeothermics. Edited by G. Buntebarth and L. Stegena. II, 234 pages. 1986.

Vol. 6: W. Ricken, Diagenetic Bedding. X, 210 pages. 1986.

Vol. 7: Mathematical and Numerical Techniques in Physical Geodesy. Edited by H. Sünkel. IX, 548 pages. 1986.

Vol. 8: Global Bio-Events. Edited by O. H. Walliser. IX, 442 pages. 1986.

Vol. 9: G. Gerdes, W. E. Krumbein, Biolaminated Deposits. IX, 183 pages. 1987.

Vol. 10: T.M. Peryt (Ed.), The Zechstein Facies in Europe. V, 272 pages. 1987.

Vol. 11: L. Landner (Ed.), Contamination of the Environment. Proceedings, 1986. VII, 190 pages.1987.

Vol. 12: S. Turner (Ed.), Applied Geodesy. VIII, 393 pages. 1987.

Vol. 13: T. M. Peryt (Ed.), Evaporite Basins. V, 188 pages. 1987.

Vol. 14: N. Cristescu, H. I. Ene (Eds.), Rock and Soil Rheology. VIII, 289 pages. 1988.

Vol. 15: V. H. Jacobshagen (Ed.), The Atlas System of Morocco. VI, 499 pages. 1988.

Vol. 16: H. Wanner, U. Siegenthaler (Eds.), Long and Short Term Variability of Climate. VII, 175 pages. 1988.

Vol. 17: H. Bahlburg, Ch. Breitkreuz, P. Giese (Eds.), The Southern Central Andes. VIII, 261 pages. 1988.

Vol. 18: N.M.S. Rock, Numerical Geology. XI, 427 pages. 1988.

Vol. 19: E. Groten, R. Strauß (Eds.), GPS-Techniques Applied to Geodesy and Surveying. XVII, 532 pages. 1988.

Vol. 20: P. Baccini (Ed.), The Landfill. IX, 439 pages. 1989.

Vol. 21: U. Förstner, Contaminated Sediments. V, 157 pages. 1989.

Vol. 22: I. I. Mueller, S. Zerbini (Eds.), The Interdisciplinary Role of Space Geodesy. XV, 300 pages. 1989.

Vol. 23: K. B. Föllmi, Evolution of the Mid-Cretaceous Triad. VII, 153 pages. 1989.

Vol. 24: B. Knipping, Basalt Intrusions in Evaporites. VI, 132 pages. 1989.

Vol. 25: F. Sansò, R. Rummel (Eds.), Theory of Satellite Geodesy and Gravity Field Theory. XII, 491 pages. 1989.

Vol. 26: R. D. Stoll, Sediment Acoustics. V, 155 pages. 1989.

Vol. 27: G.-P. Merkler, H. Militzer, H. Hötzl, H. Armbruster, J. Brauns (Eds.), Detection of Subsurface Flow Phenomena. IX, 514 pages. 1989.

Vol. 28: V. Mosbrugger, The Tree Habit in Land Plants. V, 161 pages. 1990.

Vol. 29: F. K. Brunner, C. Rizos (Eds.), Developments in Four-Dimensional Geodesy. X, 264 pages. 1990.

Vol. 30: E. G. Kauffman, O.H. Walliser (Eds.), Extinction Events in Earth History. VI, 432 pages. 1990.

Vol. 31: K.-R. Koch, Bayesian Inference with Geodetic Applications. IX, 198 pages. 1990.

Vol. 32: B. Lehmann, Metallogeny of Tin. VIII, 211 pages. 1990.

Vol. 33: B. Allard, H. Borén, A. Grimvall (Eds.), Humic Substances in the Aquatic and Terrestrial Environment. VIII, 514 pages. 1991.

Vol. 34: R. Stein, Accumulation of Organic Carbon in Marine Sediments. XIII, 217 pages. 1991.

Vol. 35: L. Håkanson, Ecometric and Dynamic Modelling. VI, 158 pages. 1991.

Vol. 36: D. Shangguan, Cellular Growth of Crystals. XV, 209 pages. 1991.

Vol. 37: A. Armanini, G. Di Silvio (Eds.), Fluvial Hydraulics of Mountain Regions. X, 468 pages. 1991.

Vol. 38: W. Smykatz-Kloss, S. St. J. Warne, Thermal Analysis in the Geosciences. XII, 379 pages. 1991.

Vol. 39: S.-E. Hjelt, Pragmatic Inversion of Geophysical Data. IX, 262 pages. 1992.

Vol. 40: S. W. Petters, Regional Geology of Africa. XXIII, 722 pages. 1991.

Vol. 41: R. Pflug, J. W. Harbaugh (Eds.), Computer Graphics in Geology. XVII, 298 pages. 1992.

Vol. 42: A. Cendrero, G. Lüttig, F. Chr. Wolff (Eds.), Planning the Use of the Earth's Surface. IX, 556 pages. 1992.

Vol. 43: N. Clauer, S. Chaudhuri (Eds.), Isotopic Signatures and Sedimentary Records. VIII, 529 pages. 1992.

Vol. 44: D. A. Edwards, Turbidity Currents: Dynamics, Deposits and Reversals. XIII, 175 pages. 1993.

Vol. 45: A. G. Herrmann, B. Knipping, Waste Disposal and Evaporites. XII, 193 pages. 1993.

Vol. 46: G. Galli, Temporal and Spatial Patterns in Carbonate Platforms. IX, 325 pages. 1993.

Vol. 47: R. L. Littke, Deposition, Diagenesis and Weathering of Organic Matter-Rich Sediments. IX, 216 pages. 1993.

Vol. 48: B. R. Roberts, Water Management in Desert Environments. XVII, 337 pages. 1993.

Vol. 49: J. F. W. Negendank, B. Zolitschka (Eds.), Paleolimnology of European Maar Lakes. IX, 513 pages. 1993.

Vol. 50: R. Rummel, F. Sansò (Eds.), Satellite Altimetry in Geodesy and Oceanography. XII, 479 pages. 1993.

Vol. 51: W. Ricken, Sedimentation as a Three-Component System. XII, 211 pages. 1993.

Vol. 52: P. Ergenzinger, K.-H. Schmidt (Eds.), Dynamics and Geomorphology of Mountain Rivers. VIII, 326 pages. 1994.

Vol. 53: F. Scherbaum, Basic Concepts in Digital Signal Processing for Seismologists. X, 158 pages. 1994.

Vol. 54: J. J. P. Zijlstra, The Sedimentology of Chalk. IX, 194 pages. 1995.

Vol. 55: J. A. Scales, Theory of Seismic Imaging. XV, 291 pages. 1995.

Vol. 56: D. Müller, D. I. Groves, Potassic Igneous Rocks and Associated Gold-Copper Mineralization. XIII, 210 pages. 1995.

Vol. 57: E. Lallier-Vergès, N.-P. Tribovillard, P. Bertrand (Eds.), Organic Matter Accumulation. VIII, 187 pages. 1995.